Crop Biosecurity

NATO Science for Peace and Security Series

This Series presents the results of scientific meetings supported under the NATO Programme: Science for Peace and Security (SPS).

The NATO SPS Programme supports meetings in the following Key Priority areas: (1) Defence Against Terrorism; (2) Countering other Threats to Security and (3) NATO, Partner and Mediterranean Dialogue Country Priorities. The types of meeting supported are generally "Advanced Study Institutes" and "Advanced Research Workshops". The NATO SPS Series collects together the results of these meetings. The meetings are co-organized by scientists from NATO countries and scientists from NATO's "Partner" or "Mediterranean Dialogue" countries. The observations and recommendations made at the meetings, as well as the contents of the volumes in the Series, reflect those of participants and contributors only; they should not necessarily be regarded as reflecting NATO views or policy.

Advanced Study Institutes (ASI) are high-level tutorial courses intended to convey the latest developments in a subject to an advanced-level audience

Advanced Research Workshops (ARW) are expert meetings where an intense but informal exchange of views at the frontiers of a subject aims at identifying directions for future action

Following a transformation of the programme in 2006 the Series has been re-named and re-organised. Recent volumes on topics not related to security, which result from meetings supported under the programme earlier, may be found in the NATO Science Series.

The Series is published by IOS Press, Amsterdam, and Springer, Dordrecht, in conjunction with the NATO Public Diplomacy Division.

Sub-Series

A.	Chemistry and Biology	Springer
B.	Physics and Biophysics	Springer
C.	Environmental Security	Springer
D.	Information and Communication Security	IOS Press
E.	Human and Societal Dynamics	IOS Press

http://www.nato.int/science
http://www.springer.com
http://www.iospress.nl

Series C: Environmental Security

Crop Biosecurity

Assuring our Global Food Supply

edited by

Maria Lodovica Gullino
University of Torino, Italy

Jacqueline Fletcher
Oklahoma State University,
Stillwater, OK, U.S.A

Abraham Gamliel
ARO Volcani Center,
Bet Dagan, Israel

and

James Peter Stack
Kansas State University,
Manhattan, KS, U.S.A

Springer

Published in cooperation with NATO Public Diplomacy Division

Proceedings of the combined NATO Collaborative Linkage Grant on
Tools for Crop Biosecurity
Torino, Italy
10–11 November 2005

Giza, Egypt
18–20 March 2006

Bet Dagan, Israel
23–27 March 2006

A C.I.P. Catalogue record for this book is available from the Library of Congress.

ISBN 978-1-4020-8477-5 (PB)
ISBN 978-1-4020-8475-1 (HB)
ISBN 978-1-4020-8476-8 (e-book)

Published by Springer,
P.O. Box 17, 3300 AA Dordrecht, The Netherlands.

www.springer.com

Printed on acid-free paper

CONTENTS

PREFACE

This book is the outcome of a Collaborative Linkage Grant on Crop Biosecurity, sponsored by NATO, which received a very positive evaluation by its Environmental Security Advisory Panel.

A legitimate question by readers could be: what has NATO to do with crop, and food safety and security? As a matter of fact, NATO is not only a political and a military organisation, but also has promoted scientific collaboration between its member and partner nations since 1958, the year in which the NATO Science Programme was established.

Nowadays, such a programme is more focused on security-related science & technology, as it was in the past, and all threats to the security and safety of populations are a matter of concern for the Organisation. This is the main reason for which in recent years NATO has sponsored a variety of scientific activities related to environmental security, such as desertification, land erosion, water scarcity and pollution, defence related environmental issues; and technical, economic and social factors underlying the management of vital resources, such as crop and food production.

Within this framework, the collaborative project on Crop Biosecurity proposed by the team of scientists lead by Professor Lodovica Gullino, was perceived by NATO as an important opportunity to increase awareness of possible threats to crop and food production caused by plant pathogens used (intentionally or not) as biological weapons.

The objectives of the project (innovative diagnostic tools; protocols for monitoring plant pathogens; eradication methods and early warning systems) were ambitious but also consistent with the qualifications of the participating research teams (which included scientists from Egypt, Israel, United States, and of course Italy).

The results of the two-year work fully met the expectations of the collaborating scientists, and also of NATO as the main sponsor of this project. As a matter of fact, I had the chance to attend one of the meetings of the different teams working on this project, and I was impressed by their competence and by their serious commitment in addressing the concepts of crop safety and security.

The quality and the quantity of the activity implemented through the NATO grant went beyond the usual standards of any Collaborative Linkage Grant traditionally sponsored by NATO. This is the main reason for which it was decided to provide additional funds in order to publish a text on this important topic.

And here is the result of the task accomplished by Lodovica Gullino and by her colleagues: a book which defines the role of crop biosecurity; outlines the consequences of plant pathogens on crop production and security; analyses the status of crop security at the local, regional and global level.

In addition, the authors also made reference to the opportunities for further collaboration and related sources for possible funding. As a consequence, this book will be a helpful tool both in becoming more acquainted with the issue of crop security, and also in being aware of the possible ways to implement further research and analysis on this subject.

In conclusion, this book reports on a success story; which is the efficient collaboration between scientists from Europe, from the Mediterranean Region and from North America, on a topic which concerns countries and societies in any part of the world and in a period of increased instability and limitation of resources.

The fact that an organisation like NATO played a role in fostering the international research activity on crop security and safety shows that these concepts are smatters of concern across the different components of modern society.

Fausto Pedrazzini
Programme Director
NATO – Science for Peace and Security

LIST OF CONTRIBUTORS

Isabelle, Bénoliel
Directorate-General
for Competition
European Commission
70, rue Joseph II, bureau 5/193
1049 Brussels
Belgium

Maria Lodovica, Gullino
Center of Competence
for the innovation in the
agro-environmental sector
(AGROINNOVA),
University of Turin,
Via Leonardo da Vinci 44
10095 Grugliasco (TO)
Italy

Will, Baldwin
Biosecurity Research Institute
1041 Pat Roberts Hall
Kansas State University
66506-7600 Manhattan
Kansas
USA

James P., Stack
Biosecurity Research Institute
1041 Pat Roberts Hall
Kansas State University
66506-7600 Manhattan
Kansas
USA

Jacqueline, Fletcher
Department of Entomology & Plant
Pathology and
National Institute for Microbial
Forensics & Food and
Agricultural Biosecurity
Oklahoma State University
74078 Stillwater
Oklahoma
USA

Federico, Tinivella
Center of Competence
for the innovation in the
agro-environmental sector
(AGROINNOVA),
University of Turin,
Via Leonardo da Vinci 44
10095 Grugliasco (TO)
Italy

Abraham, Gamliel
Laboratory for Pest management
research
Institute for Agricultural Engineering
ARO Volcani Center
P.O. Box 6
50250 Bet Dagan
Israel

CROP BIOSECURITY: DEFINITIONS AND ROLE IN FOOD SAFETY AND FOOD SECURITY

MARIA LODOVICA GULLINO
Center of Competence for the innovation in the agro-environmental sector (AGROINNOVA), University of Torino, Via Leonardo da Vinci, 44, 10095 Grugliasco (TO), Italy

JACQUELINE FLETCHER
Oklahoma State University, Stillwater, OK, USA

JAMES P. STACK
Biosecurity Research Institute, 1041, Pat Roberts Hall Kansas State University, Manhattan, KS 66506-7600, USA

Abstract: Agriculture is a key sector for the economy of many countries and it can be potentially exposed to deliberate attacks which can have dramatic economic consequences in the food, feed, and fibre sectors. Intentional introduction of certain microbes may have serious rebounds even on human health when food results contaminated after harvesting of crops and processing of raw materials. Programmes aimed at attacks against agriculture and the agro-food sector through bioweapons were conducted in the past and still in recent years this kind of threads are documented. Prevention and preparedness are the two basic approaches to maximize food security against any sort of tampering, whether natural, inadvertent or intentional. A broad and intensive collaboration among nations to strengthen the security of agricultural systems is therefore needed as in the case of the NATO funded collaborative project among EU, U.S., Israel and Egypt, which is intended to encourage greater international biodefense cooperation to protect agriculture and food systems worldwide.

Keywords: Agroterrorism; pest and disease outbreaks; accidental and deliberate introductions; response

1. Introduction

Agriculture and associated upstream and downstream related sectors are essential to the social, economic and political stability of all nations. Agricultural

M. L. Gullino et al. (eds.), *Crop Biosecurity*.

systems, including the food supply chain, are global in nature and pose a relatively soft target for those intent on harm, offering various points at which commodities could be deliberately contaminated. The threat to agriculture from exotic pathogens has long been a concern of scientists (Kingsolver et al. 1983). However, since the early 2000s, nations have become more aware of the genuine threat that terrorists or criminals could intentionally interrupt agricultural production and threaten global supplies of foods, feed and fibre (NRC 2002).

Addressing these biosecurity issues related to crops and foods has generated new fields of research (Fletcher et al. 2006, Fletcher Chapter 8 this volume) and emphasized the need for continued investment in traditional fields of plant biology. In a broad sense, plant biosecurity is aimed at protecting all plant resources (e.g., natural landscapes, agricultural lands, forests) and the food supply from the natural or intentional introduction, establishment and spread of plant pests, pathogens and noxious weeds. Securing the agro-food system requires the ability to identify high priority threats and then to prevent, detect, respond to and recover from actual events.

A number of initiatives to enhance agricultural biosecurity have been developed over the past six years in the United States (U.S.), involving many agencies and entities within the U.S. government, state and local governments, the private sector and academia (Fletcher and Stack 2007; Stack and Fletcher 2007, Stack et al. 2006). Australia and New Zealand also have invested significantly in biosecurity research, notably addressing methods to exclude, eradicate and control agents that threaten the economy, environment and human health. These new programmes focus primarily on high-consequence, existing or near-term, threatening agents and events, and often are directed toward the development of rapid, sensitive, and reliable pathogen detection technologies.

Because no nation has resources adequate to support intensive study of all plant pathogens, several U.S., Australian, and New Zealand programmes, as well as a European Union funded Crop Biosecurity project, have prioritized those agents likely to pose the greatest threats, leading to a number of lists of "most-wanted" plant and animal pathogens (Gamliel Chapter 4 this volume).

2. Agriculture as a Potential Target

The vulnerability of agriculture and the food chain to attack by those intending harm is based on several key features. One of these is the very importance of agriculture to national economies. In Europe, farming employs 5% of the workforce, and accounts for 15% of the European Union (EU) annual gross product, split nearly evenly between crops and livestock. These figures will increase when additional countries join the present EU-27. Therefore, disruption of agricultural systems could have widespread and dramatic economic consequences in the food, feed, and fibre sectors. Affected stakeholders could include farmers and input suppliers, processors, shippers, merchants, school canteens and restaurants, upstream contributors such as the agrochemical industry, and even the tourism and transportation sectors, as demonstrated recently in the United Kingdom by a devastating outbreak of foot and mouth disease. In the U.S., although farming currently employs less than 2% of the country's workforce, 16% of that workforce is involved in the greater food and fibre sector. In 2002, these industries contributed 11% to the U.S. gross annual domestic product (USDA 2004) with a total value of $1.3 trillion (USD). Agricultural exports in most years are extremely important to the balance of trade with other nations. Introduced pathogens that preclude the import of raw materials or the export of finished agricultural goods are a threat to a stable economy. Approximately 50% of all land in the U.S. (486 million hectare) is used for agriculture; 186 million hectare as cropland and 284 million hectare as forestland. In many nations, plant-based agriculture is fully integrated into animal agriculture. Impacts to plant production systems will have significant negative impacts to animal systems in the form of reduced quality of feeds (e.g., mycotoxin contamination) or reduced availability of feeds. Plant systems are the foundation of agricultural and food systems; consequently, plant biosecurity is vital to national security (Stack and Fletcher 2007).

The likelihood is very low that an attack on agricultural production could lead to starvation or significant nutritional impacts on the human population in developed nations. However, in countries having less-advanced agricultural systems an attack on a major agricultural commodity, such as rice, could lead to nutritional deprivation; even when direct consequences are less severe, developing nations are likely to recover more slowly.

Perhaps the most significant vulnerability of a nation's agricultural enterprise arises from potentially serious and wide-ranging economic impacts that could follow a deliberate attack. Most agricultural commodities are not adequately protected; for example, OECD (Organization for Economic Cooperation and Development) participating states recently recognized the vulnerability of transportation networks and are planning an experts' workshop on container security. Specific losses resulting from the intentional introduction of a virulent crop pathogen into a new location could include the value of lost production, the costs of destroying diseased or potentially diseased products, the cost of disease containment and management (vaccines, diagnostics, pesticides, and veterinary services), as well as the lost revenue to transportation and other support sectors. In addition, export markets could be lost as governments place restrictions on imported products to limit disease spread. Experience with recent foot and mouth disease outbreaks in livestock has shown that nations are quick to implement import restrictions on grounds of maintaining sanitation, but far more reluctant to remove them even after disease eradication. Multiplier effects felt more broadly might include decreased sales by agriculturally dependent businesses (farm input suppliers, food manufacturing, transportation, retail grocery, and food service) and tourism, and significant costs of compensating producers for destroyed animals or crops. Finally, because disease management regimes often call for the elimination of apparently healthy plants or livestock that might have been exposed to infectious agents, the impacts of an outbreak could extend even to producers whose commodities are not directly infected, broadening the scale of the damage even if it is contained within a small geographical region.

Other factors contributing to the vulnerability of European agriculture are the continuing trends of intensive production techniques, increasing production of genetically uniform crops and livestock, vertical integration of the production continuum, an increasing industrial dependence on the export market, and the limited presence of resistance to disease agents in key crop and livestock species. The recent threat to soybean production in the U.S., posed by a rapidly spreading soybean rust epidemic into large-scale homogeneous soybean crops, which included stockpiles of fungicides worth billions of dollars, the threat of bankruptcy for thousands of farmers, and instability of world-wide animal feed supplies, serves as a real-life example.

Finally, beyond the economic and political impacts, the ability of a terrorist or rogue state to tamper with a nation's food production systems is likely to elicit fear and anxiety among the public. This would be particularly true in the event of a public health crisis resulting from outbreaks of foodborne pathogens or the spread of zoonotic pathogens infectious to humans. Risk perception is a crucial issue in assessing the welfare losses following a terrorist attack, because effects on the demand side may exceed the losses resulting from supply disruption. Market prices of affected commodities may drop, with reductions dependent upon the degree of erosion of consumer confidence and export sales.

3. The Food Safety Issue

As agricultural products are harvested, processed, packaged, stored and transported, many become foods or food products. Intentional introduction of microbes at these points could involve plant pathogens known to reduce storage time or quality, or human pathogens such as the enteropathogens *E. coli* 0157:H7 and *Salmonella* spp. An announcement of food tampering is likely to alter consumer demands, with greater call for food products believed to be contaminant-free (i.e., customers request chicken instead of beef, or purchase foods produced elsewhere), and market prices are likely to rise for those products. Consumer confidence in their government may also be tested, depending upon the official reactions to both the threat and the eradication effort.

In general, consumers consider farms to be vulnerable targets and weak links in the farm-to-fork supply chain. Efforts to increase public awareness of both the dangers and the actions taken to prevent and prepare for agroterrorism are good ways to increase consumer confidence in the agricultural and food industry.

4. A Look at the Past

Programmes aimed at attacks against agriculture and the agro-food sector are not new, having been conducted by both nation-states and local organizations throughout history (Lederberg 1999; Madden and Wheelis 2003; National Research Council 2002; Whitby 2002). In modern times, at least nine countries (Canada, France, Germany, Iraq, Japan, South Africa, the United Kingdom, the U.S. and the former Union of Soviet Socialist Republics) had documented agricultural bioweapons programmes during some part of the 20th century. Four others (Egypt, North Korea, Rhodesia and Syria) may also have, or have had, agricultural bioweapons programmes. The agents of stem rust, Fusarium head blight of cereals, and rice blast, and the Colorado potato beetle are examples of agents that are, or were, considered for use as bioweapons. Through history there have been a number of instances, during times of military campaigns or political instability, in which civilian food supplies have been sabotaged deliberately (Kohnen 2000).

5. Potential Impacts of Exotic Pathogen Introduction

Natural outbreaks of plant disease can serve as models to estimate the destructive potential of an agroterrorist attack. Throughout history, crop diseases have contributed to significant famine. Late blight of potato swept through Ireland in 1845, ruining the country's staple food crop and contributing to the famine of 1845–46. The enormous social impact of this disease caused millions of deaths and the emigration of millions of people (Schumann 1991). In 1942–43 brown spot disease of rice contributed to the Bengal famine, in which nearly two million people died in India. Developing countries that depend on a single food crop and cannot afford imported food, are particularly vulnerable to widespread starvation following crop disease outbreaks. Given its wealth and the diversity of its food production, Europe faces little risk of nutritional hardship, but remains vulnerable to substantial financial losses (Schumann 1991).

Even a minor disease can have severe economic effects due to export restrictions. For example, in 1996 Karnal bunt was discovered in wheat seeds

·own in Arizona (U.S.) and shipped to other Southwestern U.S. states. Following the discovery of bunt spores in some seed lots, more than fifty countries – including China, the largest importer of U.S. wheat – adopted phytosanitary trade restrictions against U.S.- produced grain. Although the actual extent of this outbreak was geographically limited, the total impact on wheat exports was estimated to well over $250 million (USD). Only the highly credible, rapid and effective control and clean-up measures adopted by the affected areas minimized the impact on export trade (Kohnen 2000). Eleven years after the initial outbreak, the quarantine restrictions on affected areas are still in effect, thus limiting the economic potential of these areas.

Mycotoxins, toxic secondary metabolites produced by certain fungi, naturally contaminate a wide range of crop plants either pre- or post-harvest and are among the most important biogenic contaminants of food and feed. For example, at least 15 different *Fusarium* species colonize small grain cereals including wheat, durum wheat, triticale, barley, oats and maize, producing important mycotoxins. Some, such as aflatoxin and T2 toxin, and their fungal producers, could be attractive agents in a bioterroristic attack, leading to loss of both yield and quality. T2 toxin, sometimes known as "yellow rain", may have been in use during the Vietnam war (Tucker 2001).

With respect to insect pests, the accidental introduction into Europe of the grape phylloxera, a notorious pest of grapevine, from North America in the mid-19th century destroyed most vineyards of southern Europe with a strong impact on the economy and, in general, the whole European society at the end of the 19th century and the beginning of the 20th century.

6. More Recent Events

Deliberate contamination of food by biological agents has been reported in several countries. In 1952, the Mau Mau, an insurgent organization in Kenya, killed 34 head of cattle at a mission station using African milk bush, a local plant toxin (Kohnen 2000). In 1984, members of a religious cult contaminated salad bars in the U.S. with *Salmonella typhimurium*, causing 751 cases of salmonellosis (World Health Organization 2002). Chemical weapons have been used somewhat more commonly against agricultural targets. During the Vietnam War, the use of Agent Orange by the U.S. destroyed some crops. Other possible chemical attacks against agricultural targets include a 1997 attack by Israeli settlers, who sprayed pesticides on grapevines in two Palestinian villages, destroying up to 17,000 metric tons of grapes (Zvi 1997). In 1978, the Arab Revolutionary Council was reported to have contaminated Israeli oranges with mercury, injuring at least 12 people and reducing orange exports by 40% (Kohnen 2000).

Food production and distribution networks are susceptible to accidental on-farm contamination with human pathogens, such as pathogenic *Escherichia coli* strains and *Salmonella* species. In 1996, a disgruntled laboratory worker deliberately infected co-workers' food with *Shighella dysenteria* type 2, causing illness in 12 people (WHO 2002). Nestle reported deliberate contamination of doughnuts with *Shigella* in 1997. In 2002, 70 people were sickened by salmonella poisoning traced to a country club restaurant in Rochester, NY. The initial outbreak was followed by a second one a few weeks later; country club officials attributed the outbreaks to "deliberate contamination of food" (Isaacson et al. 2004). A recent outbreak of Hepatitis A in Pennsylvania, U.S., resulted from the consumption of green onions from Mexico: over 650 people were sickened, and three died, after eating the onions, which had been contaminated in the field by farm workers (Isaacson et al. 2004). In recent years an increase in bacterial foodborne disease outbreaks has been associated with an emerging trend for "healthy eating," increasing the popularity of salads and the consumption of uncooked vegetables (Hilborn et al. 1999). A study carried out in central Italy revealed an increase in the incidence in food borne diseases from 1996 to 2000 in private homes as well as in hospitals, canteens and restaurants (Faustini et al. 2003).

7. The Reaction

Although national governments take threats to crop health and food security very seriously, protecting agricultural commodities is challenging due to the complexity of these systems. Two basic approaches to maximizing food security against any sort of tampering, whether natural, inadvertent, or intentional, are based on prevention and preparedness (APS 2000). Prevention strategies include efforts to prevent cross-border movement of exotic pathogens, to identify and intervene in the planning of intentional introductions before such plans are implemented, and to establish physical and policy-based disincentives to agricultural attacks. Since prevention can never completely eliminate pathogen entry, it must be complemented by preparedness for the occurrence of high-impact plant diseases. This approach includes identifying high-priority pathogens in each nation, optimizing detection and diagnostic tools, understanding the biology, epidemiology and impacts of the associated diseases, developing effective management strategies for both short- and long-term, assuring robust microbial forensic technologies and skilled forensic investigators for criminal attribution, and building adequate infrastructure and communication systems to most effectively manage resources of time, funds and personnel. In addition, a comprehensive strategy must also include provisions for public information and actions to facilitate and coordinate responder networks and law enforcement agencies.

Recognition of, and effective preparation for, the threat of agroterrorism will permit governments and agrobusinesses to reduce the vulnerability of the agro-food system. Associated industries can be strengthened by the development of enhanced tools for loss mitigation, by establishing standards for, and validating diagnostic procedures, and by the certification of diagnostic laboratories.

8. The Need for an International Approach

The EU and the U.S. have well developed regulatory infrastructure for policy development and implementation that includes crisis preparedness, outbreak response and contingency plans (Bénoliel 2007; Ostfield 2007). In addition, the U.S. is strongly investing in biosecurity research, while the EU's experience and ability to coordinate food defense activities across borders, tested many times in real-life events, such as BSE and avian influenza, are valuable assets to protection efforts. Collaboration and cooperation among and between the U.S. and the nations of the EU not only will strengthen the security of agricultural systems among all direct participants, but will encourage greater international biodefense cooperation to protect agriculture and food systems worldwide.

The trends for increased agricultural trade among nations will contribute to the goal of global food security. However, this increased trade also brings heightened pressures from natural and intentional introductions of plant pathogens that threaten agricultural systems as well as food production and distribution systems. The advantages of global trade networks will outweigh the added risks if we put into place the necessary plant biosecurity infrastructure to prevent, detect, diagnose, respond and recover from disease outbreaks. The long-term costs of not developing an effective plant biosecurity infrastructure may far surpass the short-term economic benefits of global trade.

Strengthening and increasing convergence in plant biosecurity efforts across national boundaries have been the key elements in a cooperative project among the EU, U.S., Israel and Egypt. That cooperative project led to this book. Our interactions, which were enhanced by the expertise of the participating researchers, included the sharing of knowledge, views, ideas, and experiences. We explored the threat of plant pathogens to crop and food biosecurity in different regions and identified the foundations of an international plant biosecurity network. Spin-off collaborative projects in research, training, education and outreach related to agricultural biosecurity have already begun and others are planned. We invite readers to consider relevant issues in their own countries and to contact any of our project personnel for more information or to explore cooperative initiatives that will contribute to global food security.

References

APS. (2002) Crop biosecurity and countering agricultural bioterrorism: Response of the American Phytopathological Society. APSnet. http://www.apsnet.org/online/feature/bioterrorism. Accessed 15 November 2007.

Bénoliel I. (2007) EU is defending the food chain against bio attack. European Affairs 8(1): 70–74.

Faustini A, Giorgi Rossi P, Peucci CA. (2003) Outbreaks of food borne diseases in the Lazio region, Italy: the results of epidemiological field investigation. Eur. J. Epidemiol. 18: 669–702.

Fletcher J, Bender CL, Cobb WT, Gold SE, Ishimaru CE, Luster DG, Melcher UK, Murch RL, Scherm H, Seem RC, Sherwood JL, Sobral B, and Tolin SA. (2006) Plant pathogen forensics: Capabilities, needs and recommendations. Microbiol. Molec. Biol. Rev. 70:450–471.

Fletcher J, Stack JP. (2007) Agricultural biosecurity: Threats and impacts for plant pathogens. In: Institute of Medicine. 2007. Global infectious disease surveillance and detection: Assessing the challenges – finding the solutions. Workshop summary. Nat'l Acad. Press. Washington, D.C.

Hilborn ED, Mermin JH, Mshar R, Lambert-Fair MA, Farnar JA, Glynn MK, Slutsker L. (1999) A multistate outbreak of *Escherichia coli* O157:H7 contamination associated with consumption of mesclun lettuce. Arch. Intern. Med. 159:1758–1764.

Isaacson RE, Torence M, Buckley MR. (2004) Preharvest food safety and security. Am. Acad. Microbiol.

Kingsolver, CH, Melching JS, and Bromfield KR. (1983) The threat of exotic plant pathogens to agriculture in the United States. Plant Dis. 67:595–600.

Kohnen A. (2000) Responding to the threat of agroterrorism: specific recommendation for the United States Department of Agriculture. BCSIA Disc. Paper 2000–29, ESDP Disc. Paper ESDP-2000–04, John F. Kennedy Sch. Gov't, Harvard University.

Lederberger J. (ed.). (1999) Biological weapons. Limiting the threat. MIT Press, Cambridge (MA).

Madden LV, Wheelis M. (2003) The threat of plant pathogens as weapons against U.S. crops. Ann. Rev. Phytopathol. 41:155–176.

National Research Council. (2002) Countering agricultural bioterrorism. Nat'l Acad. Press. Washington, DC.

Ostfield ML. (2007) Biodefense: U.S. vision of a broader cooperation. European Affairs 8(1): 75–79.

Schumann GL. (1991) Plant diseases: their biology and social impact. Am. Phytopathol. Soc. Press, St. Paul (MN).

Stack JP, Fletcher J. (2007) Plant Biosecurity Infrastructure for Disease Surveillance and Diagnostics. pp. 95–106. In: Institute of Medicine. 2007. Global infectious disease surveillance and detection: Assessing the challenges – finding the solutions. Workshop summary. National Academy Press, Washington D.C.

Stack J, Cardwell K, Hammerschmidt R, Byrne J, Loria R, Snover-Clift K, Baldwin W, Wisler G, Beck H, Bostock R, Thomas C, Luke E. (2006) The National Plant Diagnostic Network. Plant Dis. 90:128–136.

Tucker JB. (2001) The "yellow rain" controversy: lessons for arms control compliance. Nonprolif. Rev. 8:25–42.

USDA Economic Research Service, Agricultural Outlook Tables, May 2004. www.ers.usda.gov/publications/Agoutlook/AOTables. Accessed 16 November 2007.

Wheelis M, Casagrande R, Madden L. (2002) Biological attack on agriculture: low-tech, high impact bioterrorism. Bioscience 52:569–576.

Whitby SM. (2002) Biological warfare against crops. Palgrave Macmillan Ltd., Basingtoke
World Health Organization. (2002) Terrorist threat to food. 46 pages.
Zvi S. (translator) (1997) Israeli settlers destroy 17,000 tons of grapes. Al-Ayyam, October 23, 1997.

SCOPE OF THE PROJECT

MARIA LODOVICA GULLINO
Center of Competence for the innovation in the agro-environmental sector (AGROINNOVA), University of Torino, Via Leonardo da Vinci, 44, 10095 Grugliasco (TO), Italy

Abstract: The NATO funded project "Tools for crop biosecurity" was designed to strengthen the cooperation among U.S., Europe and Israel in the field of crop biosecurity and to generate awareness on how the psychological, economic and cultural consequences of crop bioterrorism, especially attacks on soft targets such as crop seeds, could have a disproportional adverse effect on Mediterranean agriculture and, more generally, on society. The project permitted to strengthen a network of laboratories, expert on diagnostic, biology and epidemiology of pathogens and able to provide the tools for a quick diagnosis of new pathogens/race/biotypes, to study their biology and epidemiology and to share knowledge about innovative diagnostic tools for allowing a real time identification of dangerous plant pathogens. The project objectives were achieved throughout workshops and students researchers exchange.

Keywords: diagnostic tools; network; international cooperation; Mediterranean countries

The project titled "Tools for crop biosecurity" was funded by NATO within its Programme "Security through Science" and took place in the years 2005–2006. This project brought together strong scientific representation from the European Union (Italy), the United States, and Mediterranean Dialogue countries (Egypt, Israel) to develop the foundations of a multination consortium against the use of plant pathogens and pests against crops and other plant resources (forests, rangelands) as weapons. Such network has been established to improve scientific knowledge on crop biosecurity, a topic broadly addressed in the US, but less covered in EU, Israel and Egypt. This NATO project has been conceived in order to expand a network of laboratories developed within a project supported by the European Commission and actually permitted to build a stronger connection between Europe, US, Israel and Egypt.

M. L. Gullino et al. (eds.), *Crop Biosecurity*.

The project's mission has been to evaluate the nature and degree of vulnerability, the means of prevention, and the mitigation of new epidemics of plant diseases by establishing international understanding of cropping systems and products, and the development and application of new diagnostic tools. Most important was the establishment of the Eastern Hemisphere Plant Diagnostic Network (EHPDN), an interlinked network of plant diagnostic laboratories throughout Europe and the Mediterranean, in which participating laboratories established standardized diagnostic methods, shared test reagents and assessment tools, participated in collaborative, web-assisted diagnosis, and met regularly. Furthermore, the EHPDN was linked and interactive with the already-established National Plant Diagnostic Network (NPDN) of the United States (Stack et al. 2006). The project permitted sharing the "know how" among the laboratories, building strong scientific network, through a free and secure exchange, in order to develop innovative and efficient molecular diagnostic tools and to provide methods for eradication of plant pathogens. The project addressed the threat of possible use of plant pathogens as weapons against crops. The deliberate introduction of a pathogen could cause an economically disastrous disease outbreak, resulting in reductions in yield and quality, high cost disease management over the short and long term, trade disruptions and trade embargoes if certain quarantined pathogens are introduced, lost markets, higher food prices and unavailability of certain foods. Such consequences could be very difficult to be tackled in areas with limited knowledge. This project was aimed at gathering knowledge on: 1) the assessing of the risk posed to agriculture by new pathogens/races/biotypes emerging or introduced also throughout agro-terrorist attacks; 2) the prevention of the damage caused to crops by the introduction (deliberate and accidental) of new pathogens/races/biotypes; 3) the eradication and containment of plant pathogens of new introduction.

The project permitted to strengthen a network of laboratories expert on diagnostic, biology and epidemiology of pathogens and able to provide the tools for a quick diagnosis of new pathogens/races/biotypes and to study their biology and epidemiology and to share knowledge innovative diagnostic tools for allowing a real time identification of dangerous plant pathogens. A global sensibility on the issue has been achieved by strengthening technologies such as web enabled microscopy, allowing a real connection of researcher and timely confrontation on these issues. The project also permitted to share standard protocols for monitoring plant pathogens by using innovative techniques-to develop and apply molecular tools for diagnosis of known and newly introduced fungi pathogens for crops as well as containment and eradication methods for plant pathogens and early warning systems.

The project was designed also in order to generate awareness on how the psychological, economic, and cultural consequences of crop bioterrorism, especially attacks on such soft targets as crop seeds, could have a disproportional adverse effect on Mediterranean agriculture and, more generally, on society.

Project accomplishments include:

1. Establishment of the International Consortium for Biosecurity of Plants (ICBP), composed of nations participating in both the NATO and the EU projects.
2. Exchange of scientists among laboratories.
3. Workshops organized at Torino (Italy), Cairo (Egypt) and Bet Dagan (Israel), which covered scientific aspects as well as regulatory aspects. Such Workshops facilitated interaction among countries and scientists and permitted to involve national governmental officials and plant pathologists to lay the groundwork for establishment of the plant diagnostic network link in that country.
4. Common publications.

The projects, with its activities permitted to generate awareness of the problem of agroterrorism in the Mediterranean region, with a good involvement among plant pathologists and collaboration among governments. Attention was paid to training activities, needed for those with responsibilities and interest in all sectors of agriculture, including pathology extension specialists, students, crop consultants, farm advisors at various levels, regulators at national and European level.

Whereas in the United States numerous representatives of the Administration and Congress have publicly registered great concern regarding the threat of biological warfare, especially manifested bioterrorism, in Europe and in the Mediterranean agroterrorism is not yet considered at full in its threat. This project helped preparing Mediterranean agricultural system to withstand a deliberate release of a plant pathogen.

The project stressed the need for government to address, in the field of civil protection, crop bioterrorism issues and for plant pathologists to assist in developing a science-based defence plan for crops. It also urged European Union and all relevant European agencies to recognize the need to confront this threat and financially support appropriate research for fingerprinting high priority pathogens, detecting deliberate releases, developing rapid genetic-based diagnostic assays, epidemiology and risk prediction, and other scientific and technical approaches to reduce this risk.

Most researchers involved were plant pathologists by training. As already underlined, plant pathology is playing a major role, at the international level, in the study of crop biosecurity. This book relates to another one recently published in the same series (Cooper et al. 2006), dealing with plant viruses.

References

Cooper JI, Kuehne T, Polischuk VP. (eds) (2006) Virus diseases and crop biosecurity. Springer, Dordrecht.

Stack J, Cardewell K, Hammerschmidt R, Byrne J, Loria R, Snover-Clift K, Baldwin W, Wistler G, Beck H, Bostock R, Thomas C, Luke E. (2006) The national plant diagnostic network. Plant Dis. 90: 128–136.

CHALLENGES TO CROP BIOSECURITY

J. P. STACK
Biosecurity Research Institute, 1041 Pat Roberts Hall
Kansas State University, Manhattan, 66506-7600 KS, USA

Abstract: Plant systems are the foundation of food production systems and consequently, among the most important components of a sustainable society. There are many threats to plant systems that put sustainability at risk. Many challenges exist to achieving plant biosecurity at the local, regional, and global scales. Challenges to plant biosecurity include population growth, globalization, climate change, bioterrorism and biocrime, and changing agribusiness infrastructure. Additional challenges include the need for communications networks to enable collaborative diagnostics, the development of national and international technology strategies to promote the rapid global deployment of appropriate diagnostic technologies and standardized protocols, and the need for education programs regarding the importance of plant biosecurity to sustainability. It is important for each nation to develop a plant biosecurity infrastructure that ensures a safe and constant supply of food, feed, and fiber. It is equally important to develop an international framework for cooperation that maintains plant biosecurity without compromising trade.

Keywords: Agroterrorism, biocrime, plant biosecurity, crop biosecurity, communications networks

1. Introduction

Natural and agricultural plant systems are fundamental components of a sustainable society. The success or failure of many societies has been a function of the health of their plant systems (Diamond 2005). Throughout history there have been many examples of human infectious disease outbreaks leading to social unrest and political instability (Price-Smith 2002). The dynamics of impact from plant disease outbreaks can be considerably different than in human systems and the causality to political instability more subtle, but the long-term societal outcomes can be the same. Plant systems are the foundation of food production systems worldwide. In nations with lesser developed economies, plant-based foods are the primary dietary staples; when plant systems fail, there

M. L. Gullino et al. (eds.), *Crop Biosecurity*.

is no backup and food availability is decreased. In nations with fully developed economies, the diet is comprised of meat and plant-based foods. In these nations, when plant systems fail it affects the price of foods; if the effect is long-term, then ultimately the standard of living can decline. To achieve sustainability and provide for an acceptable standard of living, each nation must protect the plant resources that are consumed as food, that are fed to the animals that are consumed, that provide the fibre for clothes, the timber for shelter, and more recently, that are being used as bio-based fuels (Stack and Fletcher 2007).

The threats to plant systems are many and the probabilities of occurrence high (Gamliel Chapter 4; Fletcher and Stack 2007; Strange and Scott 2005). These threats and the risk of their occurrence have increased dramatically as a result of advancing scientific and communications technologies, increases in transportation of people, plants, and plant products, as well as the increasing risk of intentional introductions of plant pathogens to damage economies or food production capacity (Casagrande 2000; Nutter and Madden 2002). It is important that each nation develops an effective plant biosecurity plan that is compatible with its trading partners and that contributes to global food security.

The term biosecurity is used widely and in differing contexts. Often, the differences in definition are a function of the scale of the system under consideration. For the purpose of this chapter I consider security to be a state of preparedness and offer the following definitions: 1) laboratory biosecurity is a state of preparedness that ensures that specific organisms cannot escape inadvertently from a laboratory and specific organisms cannot be taken without authorization from a laboratory; 2) geographic biosecurity is a state of preparedness that ensures the containment of specific organisms within a defined area and/or the exclusion of specific organisms from a defined area; 3) plant biosecurity is a state of preparedness that protects natural and managed plant ecosystems from biological threats; 4) agricultural biosecurity is a state of preparedness that ensures a safe and constant supply of food, feed, and fibre through the prevention of, management of, and recovery from biological introductions.

2. Challenges to Crop Biosecurity

Crop biosecurity is concerned with those plants that are cultivated for the production of food, feed, fiber, and fuels. Preparedness to ensure the productivity and sustainability of these systems is confronted by a wide array of direct and indirect challenges of short or long duration. Some of the challenges are obvious in occurrence and measured impact (e.g., invasive species) while others are obscure in nature and effects (e.g., international trade agreements). Some of the challenges are independent in causality while others interactive or interdependent.

There are many specific biological threats to natural and agricultural plant systems that present challenges to crop biosecurity (Gamliel et al. Chapter 5). In addition to biological threats, there are many challenges to crop biosecurity reflected in inadequate infrastructure, insufficient international cooperation, and needed technologies. A detailed list of all the specific challenges would be too long for the constraints of this chapter. Consequently, this chapter will consider challenges to crop biosecurity in a general sense and include those that are commonly grouped into the broad categories of population growth, globalization, climate change, and bioterrorism.

2.1. POPULATION GROWTH AND DISTRIBUTION

Based on demographic trends over the last 100 years and projections for the future, increasing global populations and increasing urbanization worldwide will challenge our food production and distribution systems to satisfy the needs for safe and affordable food. Will we be able to reliably produce and distribute enough food to provide for nine billion people by 2050 and beyond? To support the tremendous increase in population we need sustainable plant systems with high productivity and high efficiency. Additional effects of expanding populations are land use changes and the encroachment on the prime arable lands utilized for the intensive agricultural production systems that support those populations.

The pressures from increased urbanization shift government and industry investments away from agriculture sectors. As economies become more developed and integrated into the global trade system, a shift from a plant-based food diet to an animal-based food diet occurs. These two factors often result in reduced funding for fundamental and applied research in plant biology and agricultural systems. It can also change the emphasis in research priorities of plant breeding programs from a focus on traits important to plant consumption by humans to a focus on traits important for plant consumption by animals.

Another challenge to crop biosecurity and ultimately food security is the emerging competition between plants for food and plants for fuel. Because the global consumption of fossil fuels may be a contributor to the current phase of climate change, alternative and renewable fuels are being developed. Plants are among the bio-based fuels currently used as replacements and supplements to fossil fuels. One consequence of the biofuel trend is the increased value of staple grain crops encouraging producers to grow plants for fuel rather than food or feed. This negatively impacts the availability and the affordability of staple foods. Also, similar to the food versus feed emphasis discussed earlier, plants for fuel result in a shift in emphasis for plant breeding programs towards traits important to processing plant materials into fuel.

At a time when sustained investment in plant research is required, the connection between people and plants is being broken. The figurative and literal distances between food producers and consumers are so great that consumer populations have very little knowledge of food production systems. Most people in developed nations have little if any understanding of the linkage between human health and plant health. Policy makers who set priorities for investments and programs are focusing on issues relevant to urban populations at the expense of policies important to sustaining agricultural systems. Rural populations have much less influence on governmental decision making processes. In addition, the short-term economic incentives for promoting international trade often overshadow the potential long-term economic and biologic ramifications of introduced pathogens and pests associated with the plants and plant products traded. The response and management costs of these introduced organisms are great. In some cases, importation can suppress local production by offering plants and plant products at lower prices than can be produced locally. Imported plants and plant products can be sold at prices lower than those produced locally ultimately suppressing local production. Consequently, when quarantines are imposed that preclude imports, local shortages can result.

2.2. GLOBALIZATION

Modern transportation systems have made possible the movement of people, plants, and plant products across oceans and continents in very short intervals of time. These natural ecological barriers (e.g., mountain ranges, oceans) to the movement of plant pests and pathogens are circumvented daily through trade and travel (Westbrooks and White 2004). To achieve cost and productivity efficiencies, agricultural and ornamental plant industries are linked across the southern and northern hemispheres to shorten production cycles and to accelerate the deployment of new plant genotypes. This is also accelerating the redistribution of pathogen species and sub-specific strains into new areas. Because of the tremendous volumes of plants and plant products in international trade, only small percentages of these are inspected at ports of entry prior to the movement of plants and plant products into new areas. This includes seeds for planting that may be colonized or infested with pathogens. It also includes ornamental and landscape plants that may be carriers (e.g., alternate hosts) of pathogens.

Another outcome of the large-scale movement of plants has been the emergence of new pathogen species made possible by bringing together species that hybridized to form new species (Brasier 2001). The global trade in landscape and ornamental plants resulted in certain *Phytophthora* species being placed in close proximity to each other. This allowed for stable interspecific hybridization leading to the evolution of a new species (Brasier 1999). Of particular concern is that the host range and the cardinal environmental conditions for disease development were not predicted by the parent species.

The consequence of this large-scale and rapid movement of plants and plant products, that are not adequately inspected, is the unintentional introduction of pathogens and insect pests into new geographic areas in timeframes that preclude preparedness. For many crop species, it still takes three to eight years to develop new commercial varieties with acceptable resistance to emerging pathogens and insect pests. Most countries lack the industrial capacity to manufacture enough pesticides (i.e., herbicides, insecticides, and fungicides) to treat most of the production areas for any single crop species. With the potential to transport a high consequence pathogen across oceans and continents in a few days, most nations are vulnerable to several years of significant losses before cost-effective management practices can be implemented over a large area.

The large-scale movement of plants and plant products among nations provides endless opportunities for bioterrorism and biocrime. Efficient yet complex transportation systems provide multiple mechanisms for introduction of pathogens and insect pests into target countries. The increasing frequency of natural introductions provides a background that will complicate distinguishing between accidental and intentional introductions precluding, in some cases, the early detection of intentional introductions. This can greatly increase the cost while lowering the effectiveness of response plans.

2.3. CLIMATE CHANGE

It is not possible at present to predict with accuracy whether the measured indicators of climate change (e.g., global warming trends, severe weather events, transitions in rainfall patterns, etc.) are of short duration or whether long-term impacts will be realized and at what scales (i.e., global, regional, or local). It is also not possible to predict with accuracy the specific impacts that climate change may have on plant diseases (Coakley 1999; Garrett 2006). However, global plant biosecurity and consequently food security will be greatly affected by climate change whether those changes are measured in years or decades. The impacts stimulated by climate change must be analyzed in the context of potential interactions with other drivers such as globalization.

The ability to predict the impact of the redistribution of species resulting from globalization is compromised by the potential impacts of climate change. Weather parameters, the regulators of plant disease development, are a function of climate. Significant climate change will impact disease development directly through changes in associated weather patterns and indirectly through changes in the species composition and development of plant communities. This in turn has the potential to result in the emergence of new pathogen species or new strains of existing species. Equally important is the potential that climate change has to provide the conditions necessary for new or existing pathogen species to be acquired by new or existing vectors. These issues will make difficult the ability to predict the potential for establishment after introduction and spread from the site of introduction.

2.4. BIOTERRORISM AND BIOCRIME

History would indicate that conflict, crime, and war have been associated with every civilization and every age. The nature of these acts, the technologies employed to carry out these acts, and the magnitude of impact have varied but the basic threats have remained over time. Progress is a function of technology. As existing technologies are improved and new technologies are developed to improve societies, the application of those same technologies and knowledge for harmful purposes has always followed. One example can be found in the communications technologies of today. Most fields from medicine to transportation have greatly benefited from the advances made in communications. Yet every day, everywhere those same communications technologies are being used for criminal purposes (e.g., identity theft, computer viruses that damage PCs or entire systems).

The vulnerability of crop production systems to deliberate attack and the ability to develop and deploy pathogens into those systems has been discussed at length (Madden and Wheelis 2003; Schaad et al. 1998; Wheelis et al. 2002; Whitby 2001). The vulnerabilities to intentional introductions of high consequence pathogens or insect pests are significant and the threats are many. Unlike acts of terrorism targeting humans or transportation infrastructure, the primary impacts of bioterrorism against plant systems will be economic in nature with the desired outcome to destabilize a nation's economy (Fletcher and Stack 2007; Kadlec 1998).

The ability to identify and prosecute those responsible for acts of bioterrorism and biocrime against plant systems will be dependent in part on the continued development of the field of forensic plant pathology (Fletcher Chapter 8 this volume, Fletcher et al. 2006). This emerging field of study needs sustained investment in order to develop the strategies and technologies necessary to distinguish intentional from accidental introductions, the ability to identify an intentional introduction that occurs coincident with a natural outbreak, and the sampling and detection protocols for multiple simultaneous pathogen introductions.

Human and animal health is dependent upon plant health. To attain global food security and sustainability, every nation must develop the plant biosecurity infrastructure that prepares for all challenges including those from bioterrorism and biocrime. Although natural and accidental introductions may pose the most urgent and frequent challenges to healthy plant systems, we must take serious the prospect of bioterrorism and biocrime. For developed nations, this may affect food affordability; for developing nations this may affect food availability.

2.5. INFRASTRUCTURE

During the last 15 years, mergers have dramatically reduced the number of companies that manufacture agrichemicals and those remaining companies have shifted their priorities away from agrichemical development and marketing to the life sciences with an emphasis on plant genetics. The discovery, development, and production of new pesticides as well as production capacity have decreased globally. In addition, improved coordination of distribution networks to facilitate the rapid deployment of fungicides to outbreak sites is needed.

Infrastructure to support secure communications and interoperability standards needs to be developed and deployed globally. To protect against deliberate attacks on food production systems and be able to identify and prosecute those responsible will require strong international cooperation among law enforcement agencies and regulatory organizations with responsibility for implementing response protocols.

3. Addressing the Challenges to Crop Biosecurity

Developing technologies and strategies for plant biosecurity will be ineffective if we lack the policies to implement them. As economies mature and standards of living rise, the distance between people and the food production systems they depend on increases. Common among many developed nations is the widescale lack of understanding of the importance of plant systems to a sustainable society, the nature and function of food production systems, and the dependence of human health upon the health of plant systems. Educating the general population and specifically policy makers is a critical challenge that must be addressed.

The very success of agriculture has resulted in policy makers that have little knowledge of food production systems and the impacts of plant diseases (Campbell 2004; Strange and Scott 2005). The comprehensive plant biosecurity infrastructure needed (Myerson and Reaser 2002) cannot be achieved without political support. Often the emphasis of policy makers is on commerce at the expense of sustainability.

Large negative economic impacts often result from the introduction of pathogens or insect pests into plant systems. In many nations, the lack of comprehensive compensation programs compromises early detection of disease outbreaks because those affected do not report the outbreak in an attempt to avoid economic losses. Policies that provide incentive for reporting outbreaks need to be enacted that cover not only the producer sector but also the distribution sectors as well.

References

Anderson PK, Cunningham AA, Patel NG, Morales FJ, Epstein PR, Daszak P. (2004) Emerging infectious diseases of plants: pathogen pollution, climate change and agrotechnology drivers. Ecol. Evol. 19: 535–44

Brasier, CM. (2001) Rapid evolution of introduced plant pathogens via interspecific hybridization. BioScience 51: 123–133

Brasier CM, Cooke DEL, Duncan JM. (1999) Origin of a new *Phytophthora* pathogen through interspecific hybridization. Proc. Natl. Acad. Sci. USA 96: 5878–5883

Burdon JJ, Thrall PH, Ericson L. (2006) The current and future dynamics of disease in plant communities. Annu. Rev. Phytopathol. 44: 19–39

Campbell FT. (2004) Political and economic barriers to scientifically based decisions. In: Biological Pollution, An emerging global menace. APS Press, St. Paul (MN)

Casagrande R. (2000) Biological terrorism targeted at agriculture: The threat to U.S. national security. The Nonproliferation Rev./Fall-Winter: pp. 92–105. http://cns.miis.edu/pubs/npr/vol07/73/73casa.pdf. Accessed 15 November 2007

Coakley SM, Scherm H, Chakraborty S. (1999) Climate change and plant disease management. Annu. Rev. Phytopathol. 37: 399–426

Diamond J. (2005) Collapse: How societies choose to succeed or fail. Viking Penguin, New York

Fletcher J, Stack JP. (2007) Agricultural biosecurity: threats and impacts for plant pathogens. pp. 86–94. In: Institute of Medicine. 2007. Global infectious disease surveillance and detection: Assessing the challenges – finding the solutions. Workshop summary. Washington, D.C: The National Academies Press

Fletcher J, Bender CL, Cobb WT, Gold SE, Ishimaru CE, Luster DG, Melcher UK, Murch RL, Scherm H, Seem RC, Sherwood JL, Sobral B, Tolin SA. (2006) Plant pathogen forensics: Capabilities, needs and recommendations. Microbiol. Molec. Biol. Rev. 70: 450–471

Garrett KA, Dendy SP, Frank EE, Rouse MN, Travers SE. (2006) Climate change effects on plant disease: Genomes to ecosystems. Annu. Rev. Phytopathol. 44: 489–509

Kadlec, Lt. Col. R. P. (1998) Biological Weapons for Waging Economic Warfare. Chapter 10 in: Schneider BR and Grinter LE (eds.) Battlefield of the Future, 21st Century Warfare Issues. Air University Press: http://www.airpower.maxwell.af.mil/airchronic les/battle/chp10. html

Madden L, Wheelis M. (2003) The threat of plant pathogens as weapons against U.S. crops. Annu. Rev. Phytopathol. 41: 155–176

Myerson LA, Reaser JK. (2002) Biosecurity: Moving toward a comprehensive approach. BioScience 52: 593–600

Nutter FW Jr, Madden LV. (2002) Plant diseases as a possible consequence of biological attack, pp. 793–818 In: Bronze MS and Greenfield RA (eds.), Biodefense: Principles and Pathogens, Horizon Bioscience, Norwich

Price-Smith AT. (2002) The health of nations. Pages 139. MIT Press. Cambridge

Schaad NW, Shaw JJ, Vidaver A, Leach J, Erlick BJ. (1999) Crop Biosecurity. APSnet Feature, September–October 1999. Online publication. http://apsnet.org/online/feature/BioSecurity/ Accessed 16 November 2007

Stack JP, Fletcher J. (2007) Plant Biosecurity Infrastructure for Disease Surveillance and Diagnostics. pp. 95–106. In: Institute of Medicine. 2007. Global infectious disease surveillance and detection: Assessing the challenges – finding the solutions. Workshop summary. Washington, D.C: The National Academies Press

Stack J, Cardwell K, Hammerschmidt R, Byrne J, Loria R, Snover-Clift K, Baldwin W, Wisler G, Beck H, Bostock R, Thomas C, Luke E. (2006) The National Plant Diagnostic Network. Plant Dis. 90: 128–136

Strange RN, Scott PR. (2005) Plant disease: A threat to global food security. Annu. Rev. Phytopathol. 43: 83–116

Westbrooks RG, White P. (2004) An ecological explosion in slow motion. In: Biological Pollution, An emerging global menace. APS Press, St. Paul (MN)

Wheelis M, Casagrande R, Madden LV. (2002) Biological attack on agriculture: Low-tech, high impact bioterrorism. Bioscience 52: 569–576

Whitby SM. (2001) The potential use of plant pathogens against crops. Microbes Infect. 3: 73–80

HIGH CONSEQUENCE PLANT PATHOGENS

ABRAHAM GAMLIEL
Laboratory for Pest Management research, Institute for Agricultural Engineering, ARO Volcani Center, P.O. Box 6, 50250 Bet Dagan, Israel

Abstract: Threatening pathogens present a great concern since they can impact agriculture, economy, international trade and environmental diversity. Among the pathogens which cause increased agricultural and social concern are new or re-emerging pathogens. Most of the important agricultural crops have spread during the last two centuries outside their original environment all over the world. An immediate and parallel trend is the migration of pathogens which accompanies the international crop spread. Thus crops become more vulnerable to new disease in their new environment. Therefore, quarantine lists for non indigenous and high consequence pathogens are specified in each country. The compilation of such lists is aimed at minimizing the invasion of these specified pests and prevents their potential damage on the local agriculture. High consequence pathogens include mycotoxin-producing organisms, bunt and smut pathogens, foliar diseases, soilborne pathogens and forest diseases. These are classified mainly according to pest risk analysis, the type of threat, type of pathogens and the circle of impact.

Keywords: Invasive pathogens, select agent list, exotic pests, crop biosecurity, emerging infectious disease, non indigenous pathogens.

1. Introduction

Virtually all plants are susceptible to a wide spectrum of diseases caused by viruses, bacteria, fungi, nematodes and other related organism sub-groups. However, the significant impact of plant disease is manifested on the horticultural crops with regard to human and animal health, food supply, and trade. Damsteegt (1999) has categorized plant pathogens into 5 main groups, according to their abundance within a specific country: (1) new – pathogens which were detected within the last five years; (2) emerging – pathogen incidence has increased within the last 20 years; (3) re-emerging – pathogens associated with chemical resistance or changes in management or cultivars, previously controlled infections;

(4) chronic/spreading – for pathogens known for longer than 20 years and causing increased concern; (5) threatening pathogens – not reported or limited in distribution in a given country. Threatening pathogens present a great concern since they can have an unexpected adverse effect on the local agriculture, but moreover on the local and regional economy, on the international trade and finally on the environment. Many terms, e.g. "invasive pests", "exotic plant pests", "emerging infectious disease (EID)", "non-indigenous plant pathogens", and other were coined in the literature to define the introduction of a pest to area beyond the boundaries of his native origin. The diversity of plant pathogens on earth together with the global changes and development create an excellent cradle for the eruption of EID, weather unintentionally or deliberately introduced. Under appropriate conditions, these pathogens can establish quickly and cause severe damages. Foxell (2001) specified the 10 major factors which contribute to the vulnerability of the current agricultural systems to the introduction of new animal and plant pests. Among these, the relevant factors to plant pathogens are: (1) the increased level of international air travel which has reduced the isolation of a country from other; (2) monoculture and the lack of crop diversity; (3) the narrow sources of plant propagation material especially, seeds; (4) the increasing reliance on pesticides to control crop pests; (5) a notable amount of imported seeds and other propagation material; (6) the lack of natural genetic immunities against the invaded species. Indeed, all the agricultural roads lead to a narrow avenue in which crop production has to flow and progress successfully, without being crushed by the increasing number of old plant pathogens.

2. Factors to Determine High Consequence Pathogens

Pest risk analysis (PRA) seems to be the main criterion to classify a pathogen as a high consequence one in the ecosystem. The most common approach to define and classify high consequence pathogens is based primarily on their relation to the crop or to food supply. Anderson et al. (2004) suggested classification of emerging pathogens into four major groups: (1) pathogens of the four staples (wheat, rice, corn and potato); (2) pathogens of cash crops and secondary food crops; (3) pathogens of non food crops; (4) pathogens of wild plants. This classification is based on global and general consideration of agriculture and food supply; however, it overlooks significant local aspects of crop systems which vary geographically and commercially among different countries and regions. PRA of pathogens which already occur in agricultural field or accidentally introduced, includes all the classical factors involved pest development, its biology and epidemiology with relation to the environment (Yang et al. 1991). In such case most of the information is available and the measures of response can be applied in order to control the damaging pathogen. Analysis of the impact of deliberately introduced pathogen brings more factors which should be considered. These factors put together the biology of the pest, its damaging impact and the ability to use it effectively as a weapon. Schaad (1999), suggest a simple numeric rating to evaluate pathogen as an agricultural weapon. He coined "Effective Pathogen Index – EPI" which is calculated from the following parameters: Toxin production, Easy to obtain, handle, and deliver; Easy to grow in large amounts; Highly infectious under many conditions; results in the establishment of a quarantine; No chemical control or host resistance available; No method for rapid or reliable detection; Infects systemically by natural means; Spreads quickly by natural means; Causes severe crop losses; Survives long periods and is persistent. Madden (2002), has proposed a population dynamic model to assess the threat of plant pathogens as biological weapons against annual crops. In 2003, Madden and Wheelis further elaborated Schaad EPI model by incorporating the threat components into Schaad scoreboard. A similar approach was proposed by Pheloung et al. (1999) for the introduction of weeds. Other approaches for Assessing the threat from invasive pest and suggesting decision-making processes were suggested also using multi-criteria methodologies (Cook and Proctor 2007).

3. List of High Consequence Pathogens

Quarantine lists for non indigenous and high consequence pathogens are specified in each country and are aimed at minimizing the invasion of the specified pest in order to prevent the potential damage to the local agriculture and economy. For example The US Animal and Plant Health Inspection Service (APHIS) have a select agent list of 8 most threatening plant pathogens (USDA 2007). The European Plant Protection Organization (EPPO) has listed two select pathogen lists: A1 – which lists pests which are absent from the EPPO region; A2 – which lists pest which are locally present in the EPPO region. In addition EPPO has an action list which aims to draw the attention of Member Countries to pests which have either recently been added to A1 or A2 lists, or present an urgent phytosanitary concern (EPPO, 2007). The quarantine lists are reviewed periodically in order to adjust them. The main intent in compiling the quarantine list is the prevention of the introduction of a selected pathogen into the country.

The quarantine regulations specify pathogens which endanger crop production of a country. A list of the most threatening plant pathogens which can be used as anti-crop weapons is not necessarily identical to the quarantine list. A list of high consequence pathogens is the first step for preparedness in order to form a quick and comprehensive response in case of a disease outbreak. Such response will rapidly overcome the pathogen's "weaponries virtues" resulting in containment of the pathogen, prevention of its spread, and eventually eradication from the invaded locus. Formal lists are available for human and animal weapon pathogens in various countries, but not for plant pathogens. Several lists were voluntarily generated and published for various purposes, such as restriction of international trade. However in most countries the only formal list is the quarantine list.

We suggest classifying the high consequence pathogens into 6 groups, based on five major factors which reflect their threat as weapons against agriculture in the EU and Mediterranean countries. The following criteria served as the guidelines for sectoring high consequence pathogens:

1. *The type of threat and the circle of its impact.*

- Food safety. Toxin production by plant pathogens in consumed agricultural products represents the highest threat, of harmful health impact on human or animal communities.
- Trade. Agricultural production in Europe and the Mediterranean countries consists of international import and export of fresh agricultural products e.g. fruits, vegetables, ornamentals, seed and other agricultural inputs and products. Thus, invasion of a high consequence pathogen pest can knock down a country's export.
- Social. The structure of many communities in Europe and the Mediterranean countries evolved around certain agricultural production systems (including production, packing and marketing). This structure is quite fragile and vulnerable to many biological threats.
- Crop and yield loss. The direct effect of a pathogen on crop and yield loss represents virtually a local damage. However, the major threat is the establishment of the pathogen in the area and its further spread and damage.
- Environment and ecological diversity. The introduction of a biological weapon to a natural habitat such as forests can alter the ecological systems, leading to negative attribute such as loss of ecotourism, recreation sites and even loss of industries such as timber wood.

2. *Type of crop.* Agricultural weapons can have the highest impact on food crops of the four staples, followed by other food crops, and non food crops.
3. *General classification of pathogen group.* The systematic grouping (i.e. viruses, bacteria, fungi), sets the type and magnitude of the agricultural threat. The nature of an organism has an important effect on the ability to prepare and spread it as anti-crop weapon, and on the ability of the pathogen to establish in a new area after it is deliberately introduced.

4. *Type of disease and disease spread.* The nature and rate of disease development and spread govern the threat potential of an introduced pathogen. The following are the main factors in assessing disease spread:

- Disease cycle. Plant disease epidemics consist of repeated cycles of pathogen development in association with the plant host and as influenced by the environmental conditions. Monocyclic pathogens produce only one cycle of development (one infection cycle) per crop cycle, while polycyclic pathogens can produce many infection cycles per crop cycle. Thus, the damaging impact from a polycyclic pathogen is potentially higher.
- Relevance of a vector. The involvement of a vector in the life cycle of the pathogen and its existence in the area of invasion, amplify the threat. The existence of the vector in or around a diseased location favors a rapid spread of infective inoculum, and rapid distribution to new infected loci. Concomitantly, the presence of a vector minimizes the prospect for effecttive containment and eradication. One of the effective spread factors of Citrus canker disease in Florida is the Asian leaf miner which is not reported in other infected countries such as Brazil and Latin America.
- Survival in the invaded area. Most (if not all) of the plant pathogens can infect other host plants even without disease symptoms. The host range of a pathogen can include plants from close botanical family, or from various cultured plants and wild plants and weeds. The host range for some pathogens (including threatening ones) can be very wide. APHIS has listed up to 97 hosts of *Phytophthora ramorum,* the causal agent of sudden oak death (APHIS 2006). Thus, the threat is bigger in such cases.
- Dissemination. A wide array of plant pathogens can infect the propagation material, mainly the seeds. The main impact of seed infection is the pathogen ability to spread over a vast area and in many locations within a short period. The transmission of a pathogen through the seeds will not necessarily express disease symptoms during the first crop generation. However it will enable the pathogen to establish in a new area long before it will be detected. Seed dissemination crosses various groups of pathogens, such as viral bacterial and fungal diseases, bunt and smuts, certain foliar diseases and others.
- Available measures for the control of invasive species. The ability to reverse the impact of an introduced pathogen and later on to contain and eradicate it from the introduced area, depends upon the available knowledge, means and technologies to maintain, eradicate and control the invading agent.

4. Classification of High Consequence Pathogens

4.1. MYCOTOXIN PRODUCING

Toxin producing microrganism fills in the top of the threat hierarchy – human and animal health. Several organisms, mainly fungi can produce toxins in the edible crop part, thereafter reaching the food. Some toxins such as deoxynivalenol (DNO), are stable even during industrial processing and can appear in processed food, thus extending the permanence of the threat. highly harmful organisms can be included in this list: Fusarium species (e.g. *Fusarium* sp, *Gibberella moniliformis, Gibberella zeae*), which attack mainly cereals and small grain; Aspergillus species (*Aspergillus flavus, Aspergillus ochraceus*), which infect a wide spectrum of plant groups including cereals, legume crops and other; Penicillium species (*Penicillium verrucosum, Penicillium solitum* var. *solitum*) which attack various cereal crops; and other (*Penicillium patulum*) which infect crops in storage.

Toxin production can be a significant threat in the EU and Mediterranean countries in many agricultural systems. Grain production and its processed products (e.g. beer and other fermented food products), are important industries in EU countries. Moreover, other vulnerable sectors include dried fruit (e.g. figs, raisins), which are important industries in the Mediterranean basin.

4.2. BUNT AND SMUT PATHOGENS

Bunt and smut diseases pose the second threat which is focused mainly on the trade of grains and meal products into and across the EU countries. The main impact of diseases such as the stinking smut (*Tilletia caries* and *T. controversa*) and Karnal bunt (*T. indica*) is by conveying bad odour, colour and taste to the grain, making it unfit for milling and resulting in the downgrade or rejection of the grains by the buyers. Typically, the level of infection includes few kernels and thus it can be easily overlooked. However the impact on the quality is huge. Some of the bunt and smut pathogens are widespread in the world. However, Karnal bunt (caused by *T. indica*) is not present yet in the EU and Mediterranean countries. Therefore, there is a great importance in limiting its spread. The bunt and smut fungi are basidiomycetes which are spread through enormous number of spores. The teliospores can survive in soil for some years.

Most of bunt and smut causing pathogens are seed-borne. Thus, the use of seeds from infected crops as biological weapons is relatively feasible and easy. Quarantine regulations and the use of resistant cultivars, can be effective in management, and control of these diseases. The use of seed treatments for the control of smuts and bunts is the best example of the effectiveness, value, usefulness, and importance of chemicals for control of cereal diseases.

4.3. FOLIAR DISEASES

High consequence foliar pathogens can cause severe yield losses. This threat can be virtually regarded as minor to the treat on food safety or on international trade. However, history shows that foliar diseases can significantly impact the demography and the economy of a nation. It is well documented that diseases such as *Phytophthora infestans* (Potato late blight) caused immigration from Ireland to the North America. In addition most of coffee production in former Sri Lanka (Ceylon) was destroyed by coffee rust caused by *Hemileia vastatrix*. Important food crops in Europe (e.g. potato, rice, and soybean) are vulnerable to devastating foliar diseases. Rice is an important crop encompassing 1.3 million hectares (mainly in Italy and Egypt). Invasion of *Magnaporthe grisea,* the fungus causing rice blast, can significantly harm rice production. Invasion of *Phytophthora infestans* to areas of potato production can result in rapid late blight epidemic. Although this fungus is widespread in Europe and routinely managed by farmers, a deliberate introduction of a new race of the pathogen, however, can change this equation. In addition, other foliar diseases which are not yet known to Europe can endanger the crops. It should be stressed however, that turning foliar pathogens to weapons, is not an obvious procedure (Whitby 2001). Nevertheless, a deliberate introduction of a pathogen in new area will take its toll following an establishment, even if after a long establishment process.

An important subgroup or foliar pathogens are bacteria which are also transmitted by vectors and pose a further threat to crops, as their spread can be extended to larger areas within a short time span. Pathogens such as *Liberobacter africanus* (the causal agent of Citrus greening), *Xanthomonas axonopodis* pv. *citri* (the causal agent of *Citrus* canker), and *Xyllela fastidiosa* (the causal agent of Pierce's disease, grapevine) are regarded as exotic pathogens. However, if these pathogens, are introduced to the EU they can spread quickly in areas were the vectors occur.

4.4. VIRAL DISEASES

The immediate threat from viral diseases is focused mainly on food crops as the damage is expressed on yield production quality. However, long-term threat which is the most significant one is derived from the nature of viruses and the lack of available control measures leading to a possible establishment of the pathogen. Most plant viruses are actively transmitted from infected to healthy plants by vectors. *Plum pox virus* (PPV) can be transmitted by at least 20 aphid species, although only 4–6 are considered the most important vectors. Aphids have been shown to carry the PPV virus on their stylets even for several kilometres. Plant viruses can have a broad host range (PPV for example infects peaches, plums, apricots, nectarines, almonds, and cherry). The existence of vectors together with wide host range makes a perfect arena for virus spread and establishment. Furthermore, viruses can be transmitted also by propagation material and parts from infected plants. Potato spindle tuber viroid can infect the tubers, can pass through both the pollen and ovule, being transmitted into seeds.

4.5. SOILBORNE PATHOGENS

Soilborne pathogens illuminate important aspects of disease threats beyond the immediate yield loss and economical damage. The main aspect of soilborne pathogens is their establishment in the introduced area and the difficulties in eradication from the soil. Once established in the area and in the soil, soilborne pathogens will continuously develop and destroy susceptible crops and further spread to new location. Other important aspect of soilborne pathogens is the wide spectrum of susceptible hosts, which further emphasizes their impact after colonization. Detection of soilborne pathogens depends upon disease symptoms, which can be delayed regarding the time of the pathogen infection. Also, a long time needed for diagnosis of some pathogens facilitates their spread and establishment before any control measures could by applied. The occurrence of Fusarium wilt of date palms "Bayoud"– a highly destructive disease – is still restricted to the western part of north Africa. The eastern parts of the Mediterranean basin, the middle East or France, comprise important date palm plantations. Invasion of the pathogen into these countries can significantly impact their economy. Thus, major efforts should focus on the prevention of the disease invasion through propagation material and other plant parts which can be transported among countries.

The threat to international trade from infected area is even more significant on a long-term basis. The introduction of soil borne diseases such as *Acidovorax avenae* sub sp. *citrulli* (*Acidovorax avenae* var. *citrulli*), the causal of bacterial blotch of melon and watermelon, can result in blocking of export of melon and watermelon fruit from a diseased area. Such restriction will apply as long as the pathogen is not eradicated from the invaded site.

4.6. FOREST PATHOGENS

Forests provide essential social functions such as recreation, and ecotourism, they provide environmental benefits and are a major factor for biodiversity conservation. Furthermore forests benefit the wood industry in Europe and the world. Pathogens of forest plants do not get much attention as they do not comply with the first circles of threat type and are not food crops. However, introduction of pathogens to forests can have a significant impact for the following reasons:

- Monitoring for pathogens in forests is not done on a routine basis as with agricultural crops;
- many forests are hard to access leaving them un-reached for monitoring, detection, or any response;
- response measures are limited;
- the involvement of many environmental factors (mountains, rivers) does not facilitate the success of any response.

Many pathogens of forest plants are restricted in specific regions in the world and do not appear in Europe. Probably, the transfer of propagation material of forests plants is smaller than that of food crops, resulting in a limited spread of pathogens. However the possible impact of such pathogens is considerable. Therefore A1 list of EPPO specifies a wide range of pathogens which pose a threat to various forest trees in Europe.

5. Disease Transmitted by Propagation Material

Dissemination is perhaps the most threatening tool for noiseless, however rapid, spread of plant pathogens. Pathogen dissemination can result in instant spread to many locations, which serve as new inoculum sources for disease eruption. The global trade of agricultural inputs and products include extensive trade in seeds and other propagation material. Seeds and other propagation material are globally traded and transported for many purposes. Agricultural industries grow crops for seed production in certain countries, in order to use them to grow the crops for food production in other, even in a different continent. Seed companies sell their seeds as the major foundation of their business. A minor, however not less important, is the transport of seeds by passengers for private purposes (Jay et al. 2003). The latter can be done by tourists looking for new exotic plants in their back yard, or immigrants seeking to maintain their traditional food in their new country.

Seeds as commercial propagation material are produced intensively in protected farms. This mainly applies to nurseries for vegetable and flower production. In contrast, the production of seeds, and other propagation material for open field crops, is done in commercial fields under commercial agriculture practices. Infection of plants and their propagation material can occur with both agricultural practices. Some examples of the possibilities for pathogen dissemination are illustrated below:

5.1. SEEDBORNE PATHOGENS

Many pathogenic viruses, bacteria and fungi can infect the seeds, or colonize the seed surface. The main dispersal path of such pathogens to long distances is by infected seeds. The spores of *Tilletia indica*, the causal of Karnal bunt can be windborne but are relatively fragile. Spread is most likely via plant parts, seeds, soil, elevators, buildings, farm equipment, tools and vehicles. In 2003, Marshall, et al., showed that at the pathway for introduction of Karnal bunt into the United States originated in Mexico. All of the interceptions of Karnal bunt of wheat into the states of Texas, Arizona, and California originnated in Mexico. Moreover, they assume that the outbreaks of the disease in central Texas were a result of long-present *T. indica* populations, which needed a buildup period to cause sufficient infections to reach detectable levels in the harvested crop. The Karnal Bunt example demonstrates also a big gap in biosecurity preparedness when it becomes to the issue of seed borne pathogens which erupt after they already establish in the area.

5.2. PATHOGENS IN SEED TUBER AND BULBS

Tubers, bulbs and other underground propagation material present an even wider dimension to pathogen dissemination. Tubers and bulbs are much bigger, and rich in nutritional components. Therefore, many pathogens may infect the tuber and reside inside the tuber (or bulb) and survive in active form, or as resting structure. Other pathogen can survive on the surface of the propagation material. Potato seed tubers are hosts for most of the soilborne pathogens of this crop. Infested potato tubers, which are imported to Israel for the spring crop from European countries are the primer source for many soilborne pathogens (Tsror et al. 1999). During the last years many pathogens were intercepted in imported seed tuber. Some of those entered resulting intensive infestation of many agricultural fields.

5.3. PATHOGENS IN OTHER VEGETATIVE PROPAGATION MATERIAL

Pathogens can be disseminated in many ways many types of vegetative material. A devastating example is the bacterium *Ralstonia solanacearum* race 3 biovar 2, that causes wilt diseases on eggplant, geraniums, potatoes, and tomatoes. The bacterium which is considered of quarantine importance in many countries may be spread by transplants and other infected propagating material, such as cuttings. Another example is the Bayoud disease of date palms which is caused by *Fusarium oxysporum* f. sp. *albedinis* and eventually results in the mortality of the infected trees. The pathogen is soilborne and its spread through soil is very limited. However, while colonizing the vascular system of the palm, the pathogen can spread by all plant parts (de la Perriere et al. 1995). Since the pathogen is confined to north west Africa, strict quarantine regulation are imposed on any propagation material of date palms from the infested area.

6. Concluding Remarks

A long list of plant pathogens is an integral part of our ecosystem biodiversity. Each crop weather food crop, forage crop, or forests is vulnerable to wide spectrum of pathogens. The origin of most pathogens is engaged with the origin of their plant host. Most of the important agricultural crops have spread during the last two centuries outside their original environment all over the world. The possibilities for diseases to follow and infect the plants wherever they currently grow is also inevitable. This trend makes the crops in their new area more vulnerable to the introduction, whether unintentional or deliberate, to plant pathogens. The reasons for the favourable conditions for the eruption and spread of high consequence pathogens were already discussed by Foxell (2001). Thus, a list of the high consequences pathogens together with all the information for preventing their invasion and the measure to control if introduced are essential for managing these threats.

References

Anderson PK, Cuningham AA, Patel NG, Morales FJ, Epstein PR, Daszak P. (2004) Emerging infectious disease of plants: pathogen pollution, climate change and agrotechnology drivers. Trends in Ecology and Evolution 19: 535–544

Cook D, Proctor W. (2007) Assessing the threat of exotic plant pests. Ecological Economics 63: 594–604

Damsteegt VD (1999) New and emerging plant viruses. APS online. http://www.apsnet.org/online/feature/newviruses/top.html. Accessed on 7 November 2007

De la Perriere RAB, Amir H, Bounaga N. (1995) Prospects for integrated control of 'bayoud' (Fusarium wilt of the date palm) in Algerian Plantations. Crop Protection 14: 227–235

European and Mediterranean Plant Protection Organization. (2007) EPPO A1 and A2 Lists of pests recommended for regulation as quarantine pests. http://www.eppo.org/QUARANTINE/quarantine.htm. Accessed on 7 November 2007.

Foxell JW. (2001) Current trends in agroterrorism (antilivestock, anticrop, and antisoil bioagricultural terrorism) and their potential impact on food security. Studies in Conflict and Terrorism 24: 107–129

Jay M, Morad M, Bell A. (2003) Biosecurity, a policy dilemma for New Zealand. Land Use Policy 20: 121–129

Madden LV (2001) What are the nonindigenous plant pathogens that threaten US crops and forests? APS Feature story: http://www.aps.org/online/feature/exotic. Accessed on 8 November 2007.

Madden LV, Bosch FV. (2002) A population-dynamics approach to assess the threat of plant pathogens as bilogtical weapons against annual crops. BioScience 52: 65–74

Marshall D, Work TT, Cavey JF. (2003) Invasion pathways of Karnal bunt of wheat into the United States. Plant Dis. 87: 999–1003

Pheloung PC, Williams PA, Halloy SR. (1999) A weed risk assessment model for use as a biosecurity tool evaluating plant introductions. Journal of Environmental management 57: 239–251

Schaad NW, Shaw JJ, Vidaver A, Leach J, Erlick BJ. (1999) Crop Biosecurity. APS Feature article. http://www. Apsnet.org/online/feature/Biosecurity. Accessed on 7 November 2007.

Tsror L, Aharon M, Erlic J. (1999) A survey of bacterial and fungal seedborne diseases in imported and domestic potato seed tubers. Phytoparasitica 27: 215–226

United States Department of Agricuture, Animal and Plant Health Inspection Service (2007) on-line: http://www.aphis.usda.gov/programs/ag_selectagent/ag_bioterr_toxinslist.html. Accessed on 8 November 2007

Wheelis M, Casagrande R, Madden LV. (2002) Biological attack on agriculture: Low tech high impact bioterrorism. BioScience 52: 569–576

Whitby SM. (2001) The potential use of plant pathogens against crops. Microbes and Infection 3: 71–80

Yang XB, Dowler WM, Royer MH. (1991) Assessing the risk and potential impact of an exotic plant disease. Plant Disease 75: 976–982

CROP BIOSECURITY: LOCAL, NATIONAL, REGIONAL AND GLOBAL PERSPECTIVES

ABRAHAM GAMLIEL
Laboratory for Pest Management research, Institute for Agricultural Engineering, ARO Volcani Center, P.O. Box 6, 50250 Bet Dagan, Israel

MARIA LODOVICA GULLINO
Center of Competence for the innovation in the agro-environmental sector (AGROINNOVA), University of Torino, Via Leonardo da Vinci, 44, 10095 Grugliasco (TO), Italy

JAMES P. STACK
Biosecurity Research Institute, 1041,Pat Roberts Hall Kansas State University, Manhattan, 66506-7600 KS, USA

Abstract: The deliberate introduction of a new plant pathogen into an agro-ecosystem could have serious negative impacts on human health, crop production and trade. Long-term effects on societies and communities surrounding the agricultural environment and agricultural industry are also possible. Invasions by new pathogens significantly increase the costs of short and long-term disease management. The production of many crops worldwide and outside their origin, together with the intensive trade of agricultural inputs and products, has increased the risk of exposure to new pathogens wherever crops are cultivated. Many of the major food staples are grown worldwide. Thus, the significance of a disease outbreak and its devastating consequences are magnified on a global scale, far beyond the region that produces that crop. The significance of crop biosecurity can increase or diminish along four major sectors (local, national, regional and global), depending upon the geographical location, the size of the area, and the nature of the agricultural system. Various aspects of crop biosecurity threats across these sectors are discussed below.

Keywords: Invasive pathogens, wheat, date-palm, tomato, potato, cacao, emerging infectious disease

1. Introduction

Agricultural crops are vulnerable to attack by a wide spectrum of plant pathogens. The deliberate introduction of a new plant pathogen into agricultural area could have a serious impact on yield and on the cost of disease management over the short and long-term. Invasions by new pathogens can disrupt trade from a country or region, if quarantined pathogens are introduced, resulting in lost markets and in some circumstances, unavailability of certain foods. The development of anti-crop weapons based on plant pathogens has been done in parallel with the development of chemical weapons and biological anti-human weapons (Wheelis et al. 2002; Whitby 2001). Although anti-crop weapons have always been of lower priority compared with anti-human ones and in most cases less successful, this threat should not be overlooked (Frischknecht 2003). Any country with the military and scientific sophistication to successfully develop a biological weapon capability is likely capable of developing an effective anti-crop biological weapon. Since certain terrorist groups and some countries are suspected of actively developing biological weapons, there is a real risk to crop systems from anti-crop weapons in today's world (Myerson 2002; Cook 2007; Desprez-Loustau et al. 2007).

The source of most plant pathogens, especially the crop-specific ones, is in most cases linked with the geographic origin of the host plant. At the site of crop origin, many pathogens formed an ecological equilibrium with the host plant and its environment (i.e. climatic conditions, natural enemies, and crop seasons) (Anderson et al. 2004). The intentional global spread of our staple food crops outside their natural environment enhances the possibilities for pathogens to redistribute and infect them at their new habitats. Crops in a new environment lack the ecologic balancing factors and are more vulnerable to severe disease outbreaks (Foxell 2001). Throughout history, outbreaks of plant diseases have been associated with crops outside their origin, resulting in wide-scale famine (e.g., late blight of potatoes in Ireland). Such consequences are more severe in developing countries, which depend on a single crop for food and cannot afford food imports. The significance of a disease outbreak within one region can be magnified on a global perspective, if that region is a major world supplier for one of the food staples. For example, *Fusarium* head blight of wheat and barley has affected several successive harvests in several states of the USA between 1993 and 1998 (Stack 2000). Such outbreaks, if deliberately initiated and further stimulated, can lead to a global food shortage.

2. Geographic Perspectives of Crop Biosecurity

Crop biosecurity can be regarded from four major geographical and area-size perspectives. Threats to crop biosecurity can be scaled up or down with respect to demography, trade issues and political concerns. The four sectors are:

- Local – a confined and relatively small area (county, municipality, geographical region) of relatively narrow spectrum of agricultural systems, usually not independent regarding phytosanitary and crop protection institutions.
- National – a country, a state, or a big province which include few or many local agricultural areas. The National level is governed by centralized crop protection and phytosanitary authorities and rules.
- Regional – a number of countries or states where mutual threats to agricultural and crop biosecurity apply. The agriculture in one country may impact the agriculture, the trade, and the economy of the countries within the region. Cooperative crop protection or phytosanitary authorities may exist.
- Global – independent and interdependent crop production and trade systems of global scale. This is the largest spectrum of crops and an almost infinite reservoir of plant pathogens, with the potential to invade and attack any available niche.

The parameters that are relevant to crop biosecurity perspectives that will be discussed in this chapter are:

- *Area: size, location and boundaries.* The geographic location, topographical contours size of the area, and agricultural and political neighbours.
- *Crops, production systems and markets.* The type and spectrum of crops which are grown in the area, production systems, e.g., food crops, intensive greenhouse production, open field production, etc. An important factor is the end result of the production (local market, industry, fresh product export).
- *Possible and visible threats.* The possible plant pathogens which can harm the crop systems, including indigenous and exotic pathogens. The major threats concerns are: (1) Food safety from toxin producing pathogens; (2) disruption of trade; (3) loss of crop yield; and (4) disturbance of the environmental and ecological biodiversity.

- *Preparedness (organization, detection, diagnosis, response).* The existence and structure of disease detection and diagnosis networks in the field and in a central organization; the establishment of crop protection and phytosanitary authorities; the enforcement of local and central regulations regarding quarantine and prevention; and the availability of plans, facilities and measures for response.

3. The Local Perspective

Area and boundaries – The local perspective of biosecurity refers to a confined geographical area with specific topographical contours and climatic conditions. This may be a region with defined borders which separates it from other regions. In most cases a local area is located in proximity to other agricultural areas and also in the vicinity of a populated urban area.

Crops, production systems and markets – Crop diversity in a local area is usually narrow, especially in intensive agriculture which focuses on a few crops. Some specific productions systems are intended for specific markets (e.g., export, local food consumption, grains, industrial processing, etc.). The specialization in specific production systems governs specific procedures which are relevant to crop biosecurity, such as intensive import of propagation material and the use of recycled packing material (e.g., boxes). On the other hand, a local area can contain a vast array of crops and production systems mainly for local consumption. This is often the case for developing countries, in which a local community supplies most of their needs.

Possible and visible threats – Threats to crop biosecurity from the local perspective are usually specific to the relevant crops. The main concern is for plant pathogens which attack these crops and which do not exist yet in this specific location. The threatening pathogens should include not only exotic quarantine pests, but also pathogens that are present nationally but are not in the specific local area. The ramification of the last scenario is significant when the pathogens appear in neighbouring agricultural areas.

Preparedness – Local phytosanitary and crop protection system can be based on national rules and local initiatives. This can include local quarantine measures and local detection and surveillance network.

3.1. ISRAEL – ARAVA

Area and boundaries – The Arava Valley is the southern part of the Syrian-African rift: it is located along an area from the Dead Sea to the Red Sea and is surrounded by mountain ranges along both sides. The population and the agricultural system consist of small agricultural communities which are scattered along the valley.

Crops, production systems and markets – The climatic and soil conditions in the Arava area permit a narrow spectrum of crop and cropping systems. The main crops grown in the area are tomatoes, peppers, and melons for fresh market export during the winter. Intensive agriculture is practiced which includes monoculture of more than one crop during the year, soil fumigation to eliminate soilborne pathogens, and pest management. The main plantation fruit is date palms which are grown mainly for export. The best quality cultivars are grown in this area. Date palms are vulnerable to some quarantine pests which can significantly damage this production system.

Possible and visible threats – Invasion by new pathogens of the main crops in the Arava valley have already occurred in the past. However, many pathogens are still regarded as quarantine pests or do not exist in the area. *Acidovorax avenae*, a bacterium causing melon and watermelon fruit blotch, can disrupt the export market if enters and erupt in the area. Similarly, Phytophthora blight, (caused by *Phytophthora capsici*) another quarantine disease is a threat to pepper and other solanaceous crops. Fusarium wilt of date palms (Bayoud) is currently contained in the western part of north Africa. This pathogen is a major threat to the date industry if spread to the Middle East. However, currently the major threats on tomato production in the Arava valley are the pathogens, *Phytophthora infestans* and *Clavibacter michiganensis*. These pathogens are widespread in Israel and cause substantial damages annually. Since the two pathogens exist in Israel, they do not appear in the quarantine list of pathogens; however their threat is more visible.

Preparedness – Since the Arava is export-producing; it is recognized by the Israeli Plant Protection and Inspection Service as a quarantine zone. A quarantine zone has specific regulations and procedures for the transfer of agricultural products from and into the Arava valley (Aahwal 2007). In addition, farmers are required to implement a four week "no crop" period for the entire area during the summer. The goal of this regulation is to prevent the full life cycle of certain pests and to suppress their establishment in the fields. Agriculture in the Arava features a phytosanitary and crop protection system consisting of pest monitoring scouts and pest control advisors.

3.2. ISRAEL – WESTERN NEGEVE

Area and boundaries – The western Negev is the major agricultural part of southern Israel. The topographic borders are not obvious. It is a flat area that borders the Gaza strip in the west and the Sinai desert in the south. The population consists of villages, small urban communities and towns. The agricultural production system consists of small as well as big farms.

Crops, production systems and markets – A wide spectrum of crops and cropping systems are being employed in the area. The main crops are potatoes, tomatoes, peanuts, wheat, carrots and many more. Most of the crops are grown for export as well as domestic consumption. Open field farms practice intensive agriculture with narrow crop rotations on each farm. On the small farms, greenhouse production consists of intensive monoculture practice.

Possible and visible threats – The main concern for pathogen invasion is through potato seed tubers which are imported from western Europe and are the propagation material for the spring crop. The vicinity of many related crops, together with the wide spectrum of agricultural systems and urban areas makes crop biosecurity a difficult task. Furthermore, the proximity to the intensive agricultural area of the Gaza strip together with the lack of cooperation between the two areas facilitates pathogen movement to new areas. Thus, the invasion of known pathogens, such as new races of *Phytophthora infestans* and *Clavibacter michiganense*, are very feasible. Being a major potato production area, potato seed tubers are imported for the spring season from Europe. The imported seed tubers can serve as a means for the introduction of certain high consequence pathogens such as *Synchytrium endobioticum*, the causal agent of potato wart and *Clavibacter michiganensis* subsp. *sepedonicus*, the causal agent of potato ring rot. Invasion of pathogens through imported seed tubers has occurred in the past (Tsror et al. 1999): thus, it poses a very real threat. Invasion of quarantine pathogens such as *Phytophthora capsici* (cause of Phytophthora blight), and other pathogens is a real threat by means of infected seeds and other random or deliberate actions.

Preparedness – There are no special regulations for this area by the Israeli Plant protection and Inspection Service and there are no local regulations. The agricultural phytosanitary and crop protection system consists of individual farm initiatives based on market contracts. These are operated by pest monitoring scouts and pest control advisors on a local basis.

3.3. ITALY – PIEDMONT

Area and boundaries – Piedmont is a large cultivated area located in north-western Italy. The agricultural production system consists of small to large farms.

Crops, production systems and markets – Agricultural production in Piedmont is quite important and accounts for a significant percentage of the national total. There is a wide range of crops and cropping systems: cereals (421,467 ha), grapevine (53,328 ha), fruit (apple, peach, hazelnut) (31,273 ha) are among the most popular crops. Rice is one of the most important crops, covering 117,835 ha, accounting for 50% of Italian production. Crops in this region are grown for domestic consumption as well for export. Integrated production systems are quite widespread and a good extension service is available to growers.

Possible and visible threats – The main concern for pathogen invasion is through seeds and propagative material. Recent outbreaks of *Aphelencoides besseyi* on rice, of *Dryocosmus kuriphilus* on chestnut, of *Diabrotica virgifera* on corn, of *Phytophthora ramorum* on azalea are good examples of new pest and pathogen introductions that threaten the crops. The proximity of several different crops, the broad spectrum of crop systems adopted, the close connection of farms and cities is a challenge for crop biosecurity.

Preparedness – There are no special regulations for this region nor local rules concerning crop biosecurity. In most cases, the agricultural phytosanitary and crop protection approaches follow the integrated production strategy, with relatively good pest monitoring scouts and pest control advisors evenly distributed throughout the territory.

3.4. USA – KANSAS

Area and boundaries – The State of Kansas is centrally located within the Great Plains region of North America and covers approximately 210,000 square kilometres. The climate within Kansas varies along a southeast to northwest transect and includes three eco-regions. Rainfall varies across this area from over 112 centimetres annually in the east to less than 30 centimeters annually in the west with an elevation gradient of 200 meters above sea level in the southeast to over 1000 meters in the northwest. This variation results in 200 growing days in the southeast to 150 growing days in the northwest. Kansas is bordered by tall grass prairie on the east high desert plateau on the west.

Crops, production systems, and markets – With respect to natural and agricultural plant systems, Kansas is a microcosm of the Great Plains Region. Consequently the cultivated areas are dominated by wheat (ca. 4 million hectares), maize (ca. 1.2 million hectares), sorghum (ca. 1 million hectares), and soybeans (ca. 1.4 million hectares). Intensive agricultural production practices dominate this area and include an array of crop rotation schemes and cultivation practices. Long-term monoculture (e.g., over 50 years of continuous wheat or maize) is common and may be adjacent to integrated production systems practicing minimum tillage with multiple crop rotation patterns (e.g., wheat-soybean-fallow-maize). In Kansas, plant-based agricultural systems are fully integrated with animal-based agricultural systems that are highly developed and productive. Consequently, the animal production systems are dependent upon plant based agriculture for feed, forage, and grazing. The high efficiency of crop production systems in Kansas, as well as the Great Plains Region as a whole, results in high quality feed grains at low cost. Consequently, food animals (e.g., cattle and swine) are transported into Kansas to complete the animal production cycle. In some years, over 55% of the wheat produced is exported internationally. Maize, sorghum, and to some extent soybean may also be exported. This includes grains for animal feed, grain for human consumption, as well as seed for crop production in other nations.

Possible and visible threats – As stated previously, the climate and elevation vary dramatically across Kansas resulting in extremely variable weather patterns that impact disease development and spread. Planting dates for the major crops are sequential going from south to north. Kansas is central located in the Great Plains Region. As such, during the long growing season it is subject to weather fronts from the northwest, southwest, and southeast. The consequence of this sequential planting along with the varying wind directions during the growing season facilitates the spread of pathogens and insect pests from disease outbreaks anywhere in the Great Plains Region.

Kansas is located in the middle of the "Puccinia Pathway" (Brown et al. 2002). Recent epidemics of stripe rust (*Puccinia striiformis*) and leaf rust (*Puccinia recondita*) of wheat and southern rust of maize (*Puccinia polysora*) have caused very significant annual losses. In most years up 50% of the wheat varieties are planted to varieties with common genetic backgrounds that are susceptible to the leaf and stripe rust pathogens. The consequence of this is reflected in the approximately $250 million (USD) losses to three stripe rust epidemics over the last six years. The emergence of new races of the stripe rust pathogen as well as the new Ug99 race of stem rust of wheat (*Puccinia graminis* f. sp. *tritici*) that emerged in eastern Africa are of great concern. Most wheat varieties are susceptible to Ug99; if it gets introduced into the southern Great Plains, the potential to spread across Kansas and cause very severe losses is significant.

Preparedness – Large-scale and small-scale disease surveillance systems are present in Kansas. The State Plant Health Director and the State Plant Regulatory Officer ensure enforcement of quarantine regulations and inspecttions of plant trade across state borders. Researchers and extension personnel at Kansas State University (K-State) manage local disease monitoring systems, often in partnership with government agencies, agricultural industries, and crop advisors. The K-State Cooperative Extension system ensures the dissemination of information regarding disease development and disease management options for producers. Kansas is a member state of the Pest Information Platform for Extension and Education (PIPE), a national disease surveillance system comprised of an internet-based reporting system and a sentinel plot monitoring system for Asian soybean rust and legume viruses (Isard et al. 2006). K-State houses the regional centre laboratory for the National Plant Diagnostic Network (NPDN). NPDN is a national network of university, government, and industry plant diagnostic laboratories with the capability for collaborative internet-based diagnostics, training, and secure communications (Stack et al. 2006).

4. The National Perspective

Area and boundaries – The national perspective of biosecurity refers to the political boundaries of the country (or a state in federation such as USA, Canada and other). The topographical contours of a country are usually variable as are the climatic conditions. The national perspective always involves the interaction with the population and plant issues which are related to it. The main interaction and friction points are the limitations in pest control, and the free move of anything among the various regions of a country.

Crops, production systems and markets – A country can have crop diversity in different locations where each location may focus on few crops. However, there are countries or states in which few crops are grown on a commercial scale. There are various productions systems which intend for wide markets, local and export. The variety of production systems involves many factors which are relevant to crop biosecurity, such as distribution of propagation material and plant related materials (e.g., boxes). On field crops this involves the move of contractor equipment and machinery across the country with all the consequences of pathogen spread (e.g., crop harvesters); in developing countries where agriculture is based on small farms these aspects are minor in importance.

Possible and visible threats – The threats to crop biosecurity from the national perspective are wide on one hand according to the wide spectrum of the crops. However the main threat is plant pathogens which attack the major crops such as the food staple crops and those which comprise the backbone of the agriculture such as export. The threatening pathogens appear in national list of quarantine pests, special concern should be taken regarding the quarantine pests which appear in neighbouring nations, as they do not obey political boundaries.

Preparedness – Each state or country usually has a Governmental plant protection and inspection unit, which is the formal authority for phytosanitary and crop protection. Other institute which exits in conjunction or separate is central diagnostic institute. Each state forms the regulation for quarantine, prevention of pathogen invasion and for response.

4.1. ISRAEL

Area and boundaries – The Israeli perspective of biosecurity refers to two major factors: the long borders, and the political boundaries with Syria, and Lebanon of which there are no direct relationship. The agriculture in Israel involves the interaction with the population as some of the intensive agricultural areas (especially in central Israel) are located within urban and populated area.

Crops, production systems and markets – Being a subtropical country with various climatic conditions a wide spectrum of crops are grown in different locations where each location may focus on few crops. Many vegetable and fruit crops are grown under intensive cultivation for export. The major export crops are potatoes, vegetables (tomatoes, peppers, fresh herbs) and others. The variety of production systems includes greenhouse production as well as row crops in open fields and extensive field crops.

Possible and visible threats – The threats to crop biosecurity from the Israeli perspective are wide according to the wide spectrum of the crops. However the main threat is plant pathogens which attack the major export crops. Among the major threats potato pathogens are *Ralstonia solancearum* the causal agent of potato brown rot; *Synchytrium endobioticum* the causal agent of potato wart; *Clavibacter michiganensis* subsp. *sepedonicus* the causal agent of potato ring rot. Vegetable production is threatened by a number of quarantine pathogens such *Phytophthora capsici* the causal of Phytophthora blight, of peppers. The major threat on date palm production is Fusarium wilt of date palms (Bayoud) which is currently contained in the western part of north Africa. The threatening pathogens appear in national list of quarantine pests, special concern should be taken regarding the quarantine pests which appear in neighbouring nations, as they do not obey political boundaries.

Preparedness – The authority for crop biosecurity is the plant protection and inspection service (PPIS). There is a quarantine list of selected agents which is regularly updated. Upon detection of a quarantine pest, a routine of eradication procedure are enforced and executed.

4.2. ITALY

Area and boundaries – The Italian perspective of biosecurity mainly refers to its long sea borders, which permit a very easy and often uncontrolled arrival of people and goods. Another important aspect is linked to the intensive cropping systems often adopted and with the relative importance of the vegetable and ornamental sectors, whose dependence on imported propagation material is well known.

Crops, production systems and markets – Italy is a subtropical country with various climatic conditions and a wide spectrum of crop grown in different areas. Cereal crops cover 3,801,047 ha; potato is grown on 72,451 ha, open field grown tomato covers 114,563 ha (80% is grown for processing). Greenhouse crops are important in central and southern Italy. Vegetable, ornamental and fruit crops are grown very intensively and are often exported. The major export crops are vegetables (tomato, pepper, eggplant, fresh herbs), fruit (apple, peach, kiwi) and others.

Possible and visible threats – The threats to crop biosecurity from the Italian perspective are wide according to the wide spectrum of the crops grown under intensive cultivation systems. The main threats are represented by a number of pathogens which can be introduced throughout infected seeds of propagation material. The most threatening pathogens appear in national list of quarantine pests, special concern should be taken regarding the quarantine pests which appear in neighbouring countries. Beside quarantine pests, many new diseases appeared recently on economically important crops. Most of them were imported through infected seeds and propagation material.

Preparedness – The authority for crop biosecurity in Italy is the plant protection and inspection service. The EPPO quarantine lists are followed and updated.

4.3. USA – NATIONAL

Area and boundaries – The United States is a vast area stretching from Mexico in the south to Canada in the north and bordered by the Atlantic Ocean on the east and the Pacific Ocean on the west; approximately 3000 kilometres south to north (25° N to 45° N latitude) and 6720 kilometres from east to west (68° W to 168° W longitude). The USA includes the Hawaiian Islands and Alaska that are not contiguous with most of the nation. The national climate varies from subtropical to temperate and rainforest to desert. Four major mountain ranges run south to north across the country. The USA includes many ecological regions and many habitat types that are important as reservoirs of pests and pathogens.

Crops, production systems, and markets – The USA has a very broad array of simple and complex crop systems that provide raw materials and finished agricultural products to local, regional, national and international markets. These include many of the major grain crops (e.g., wheat, maize, sorghum, rice, barley, oats), oil seed crops (e.g., soybean, peanut, sunflower), fruit crops (e.g., citrus, pome, stone, berries), nuts (e.g., walnut, pecans), vegetables (e.g., cole crops, root and tuber crops), forest (e.g., natural systems and plantation hard and soft wood species), and a broad array of ornamental plant systems.

Possible and visible threats – Due to the very large geographic scale of the USA, climate and elevation vary dramatically across the nation. Some crops are planted across very wide areas such as wheat, maize, beans, apple, citrus; sometimes covering the vast expanse of coast to coast. This creates the risk of severe negative impacts (e.g., economic losses, ecological destabilization) from introduced pathogens or insect pests with the ability to spread very long distances. Agriculture and agricultural exports are an important component of the US economy with an annual value of $1.3 trillion (USD).

Several lists have been compiled by different agencies and organizations of pathogens that pose a significant threat to US agricultural systems. By law, the US Department of Agriculture (USDA) is required to develop a "Select Agent List" that identifies the most threatening plant pathogens to the USA that could be intentionally used (Fletcher and Stack 2007). Different criteria are used to generate those lists, consequently there are many pathogens identified as threatening. A few of the pathogens included in those lists are: *Puccinia graminis* f. sp. *tritici* race Ug99 (stem rust of wheat), *Ralstonia solanacearum* race 3 biovar 2 (brown rot of potato and tomato), *Liberobacter americanus* (Huanglongbing, or citrus greening), and *Phytophthora kernoviae* (rhododendron).

Preparedness – Preparedness for the natural, accidental, or intentional introduction of plant pathogens into the USA is achieved through a broad spectrum of narrow-focus local programs to very complex multi partner programs covering large areas and multiple crop species. These partnerships involved public and private sector entities with varying levels of authority and responsibility for the target system. The US Department of Homeland Security Office of Customs and Border Protection is tasked with intercepting potentially destructive pathogen and pests at ports of entry. The US Department of Agriculture is tasked with managing off-shore disease monitoring programs, domestic outbreaks of introduced pathogens, as well as interstate movement of plants, plant products, and pathogens. Universities partner with government agencies to develop research, education, and outreach programs regarding crop diseases. The American Phytopathological Society contributes to setting national priorities for research and education (Schaad et al. 1999).

Large-scale and small-scale plant disease surveillance systems are present across the USA (e.g., Main et al. 2001); some are described in the other sections. The National Plant Diagnostic Network (NPDN) is a partnership among government and university institutions with disease diagnostics and surveillance capability (Stack et al. 2006). An area in need of attention is the education of the general public. The importance of plants to human health and to economic well being of society is not widely understood. The tremendous increase in internet trade in plants and plant products and the rapid long distance travel of very large numbers of people across the globe have significantly increased the risk of accidental introduction of pathogens. The role of the average citizen in plant protection is important.

5. The Regional Perspective

Area and boundaries – A regional perspective of biosecurity refers to a group of countries which are geographically related such as the Middle East, the Mediterranean basin and Europe. A region can be referred also to a federation such as USA, Canada and the European Community. A region is characterized in many cases by open borders and free move of people plants and plant products. In some cases however, hostility among neighbouring countries keep close borders along with lack of knowledge transfer, also with regard to pests and pest threats.

Crops, production systems and markets – a region has wide crop diversity and crop production systems. However, there are some principal crops which are important to most countries in the region. For example date palms are an important export crop in North Africa and the Middle East countries. Commercial treaties among countries in the region include trade in fruits and plant products. This involves also propagation material and artificial substrates which can contain invasive pathogens. Seeds are one of the most important trades among countries.

Possible and visible threats – The threats to crop biosecurity from the regional perspective are wide on one hand according to the wide spectrum of the crops. The threats are divided into two main groups. The first is pathogens which do not appear in any of the countries within the region. The second is pathogens which appear in some of the countries but other countries are free. The threats of the latter are more realistic when the boundaries between countries are close and so is the knowledge.

Preparedness – Regions which are organized such as federation or the EU have usually formal organizations plant protection and inspection unit, which is the formal authority for phytosanitary and crop protection (such as the European and Mediterranean Plant Protection Organization, EPPO). Such institute usually works in harmony with the national authorities to form the procedures and regulation for quarantine, prevention of pathogen invasion and for response.

5.1. THE MIDDLE EAST

Area and boundaries – The middle East perspective of biosecurity refers to three major components. (a) long coastal strips of the Mediterranean, the Red sea and the Persian gulf, along which a significant part of the agriculture is located and vulnerable to invasion of new imported pathogens; (b) the vast inland remote areas in which intensive and extensive agriculture is practiced, however scouting and monitoring is not frequent; (c) the hostility among neighbouring countries and an unstable situation in others (Iraq) prevents knowledge transfer and cooperation with regard to pests and pest threats.

Crops, production systems and markets – The middle East has wide crop diversity and crop production systems, from cereal (rice) in Egypt to intensive greenhouse production along the Syrian African rift (Israel, Jordan, Lebanon) and Turkey. However, there are some staple crops which are important to most countries in the region. Wheat and cereals are grown in most of the countries and are susceptible to migration of spores from infected areas. Another crop, date palms, is an important export crop from many Middle East countries (Egypt, Iraq, Jordan, and Israel). Commercial treaties among countries such as Israel, Egypt and Jordan, include trade of fruits and plant products. This involves also propagation material which can spread destructive pests such as Bayoud (*F. oxysporum* f. sp. *albedinis*) and the palm red weevil.

Possible and visible threats – The threats to crop biosecurity in the Middle East region are as wide as the length of the coast and wide as spectrum of the crops. The first major threat is new diseases. For example the Ug99 strain of wheat stem rust, that was isolated in Uganda during 1999. The pathogen is virulent to most wheat cultivars and its predicted migration path is to North Africa through Arabian Peninsula and then to Middle East and Asia. The second is pathogens which appear in some of the countries but other countries but not in other. Karnal bunt of wheat (*Tilletia indica*) appears in the eastern skirts of the area (Iran and Afghanistan) and poses a threat to all the wheat production areas (EPPO 2007). This threat is more real as the infected area is also politically unstable making knowledge transfer difficult. White rot on onion (*Sclerotinia cepivorum*), is well established in Egypt and could cause severe damage to the Eastern basin of the Mediterranean if arrives. The threats of the latter are more realistic when the boundaries between some countries sharing the same threats are closes and so is the knowledge transfer.

Preparedness – Unfortunately, the Middle East has no organized or formal organizations plant protection and inspection unit for phytosanitary and crop protection. The intra-region cooperation involves joint work between some countries (e.g., joint monitoring of date palm pest in Egypt, Israel and Jordan). However infrastructure for a region-wide cooperation is absent and critically needed.

5.2. THE MEDITERRANEAN BASIN

Area and boundaries – The Mediterranean basin encompasses countries from three continents, all sharing one big harbour – the Mediterranean Sea. This region is characterized by a mild climate which allows similar agricultural systems in many countries in the regions. In addition the mutual conditions and cropping systems in the region makes the agriculture vulnerable to the invasion of new pathogen. In contrast, the countries around the Mediterranean Sea belong to separate geopolitical regions with different national and semi-regional political and commercial aspiration, which in turn, may conflict with the regional crop biosecurity needs.

Crops, production systems and markets – The agriculture around the Mediterranean basin consists of wide crop diversity and crop production systems, from extensive production of staple crops (wheat rice, potatoes), intensive vegetable and floriculture production in greenhouse, and fruit tree production. The staple food crops i.e. wheat and potatoes, are of mutual interest to most of the countries in the region. In addition, intensive vegetable production is an important agricultural industry for local and export (mainly Europe). The trade of vegetables fruits and flowers, from the Mediterranean basin to Europe consist of the market demand for stable and constant supply of these products. Thus, the transport of agricultural productions within The Mediterranean basin and Europe is constant. For example, a major part of pepper consumption in Europe comes from Spain, Italy and Israel. Some crops however are specific to few countries in the region. Date palms, is an important export crop from the Middle East and North Africa (Israel Egypt, Libya, Algeria Tunisia and Morocco).

Possible and visible threats – There are many threats to crop biosecurity in the Mediterranean basin which are derived from similarity in crops and production systems, the mutual markets and trades, and the vulnerability of the main crops in the region. Potato seed tubers are traded within countries in the Mediterranean basin. This can serve as a mean for the introduction of certain pathogens such as *Synchytrium endobioticum* the causal agent of potato wart and *Clavibacter michiganensis* subsp. *sepedonicus* the causal agent of potato ring rot, exist in Europe and in Turkey. Thus, these pathogens pose a real threat to the potato production around the Mediterranean basin. Another example of a pathogen is Fusarium Bayoud of date palms which is confined to Morocco and Algeria. Migration of this pathogen further east can adversely impact the date production and export, especially in intensive cultivation system such as Israel. A wide spectrum of new pathogen is visible threat to the staple crops in the Mediterranean basin. Early mentioned are the Ug99 strain of wheat stem rust, and Karnal bunt of wheat (*Tilletia indica*) which already infect areas close to the Mediterranean region.

Preparedness – The Mediterranean basin as the Middle East has no organized or formal organizations for plant protection and inspection unit for phytosanitary and crop protection. Some sectors, i.e. the European part work according to EPPO regulation, a scientific organization, The Mediterranean Phytopathological Union (MPU), serves as a scientific forum for the exchange of information among scientists. This organization, however, is not regional organization for cooperation which is needed.

5.3. EUROPE

Area and boundaries – The European Union comprises countries from various geographical and agricultural sectors, the Mediterranean Sea, North Central and Eastern Europe. Europe is characterized by variable climatic conditions in the different zones. Thus, a wide array of agricultural systems exists in the countries in the regions. In addition the mutual conditions and cropping systems in the region make the agriculture vulnerable to the invasion of new pathogen. Most of the countries in the European continent belong now to the European Union (EU) although the agricultural nature and commercial aspiration of different countries in the union are different and thus may consider different crop biosecurity needs.

Crops, production systems and markets – The agriculture in Europe and in the Mediterranean basin consists of wide crop diversity and crop production systems, from extensive production of staple crops (wheat rice, potatoes), intensive vegetable and floriculture production in greenhouse, and fruit tree production. The staple food crops i.e. wheat and potatoes, are of mutual interest to most of the countries in Europe. These crops are grown mainly in the large countries on large field scale. On the other hand, intensive vegetable production distinguishes agriculture in southern Europe and the Mediterranean countries. The southern part serves as the vegetable basket of Europe. Thus, the trade of vegetables fruits and flowers, from the Mediterranean countries of Europe consists of short time of transport. The agriculture in the central and Nordic countries focuses on arable crops, deciduous fruit trees, and also greenhouse production.

Possible and visible threats – There are many threats to crop biosecurity in Europe. These derive first from the specific threat of each sub-region (i.e. the Mediterranean countries, Central Europe, and northern countries). Potatoes are very vulnerable to pathogens which do not yet exist in Europe. For example the Mediterranean countries are still free from the pathogens *Synchytrium endobioticum* the causal agent of potato wart and *Clavibacter michiganensis* subsp. *sepedonicus* the causal agent of potato ring rot, which exist in Northern Europe. A wide spectrum of new pathogen is visible threat to the main staple crops in Europe. Early mentioned Ug99 strain of wheat stem rust and Karnal bunt of wheat (*Tilletia indica*) are still absent in Europe.

Preparedness – The formal organization for plant protection and inspection unit for phytosanitary and crop protection, is EPPO regulation, a scientific organization.

5.4. USA – GREAT PLAINS REGION

Area and boundaries – The Great Plains Region of the United States is a vast area stretching from Mexico in the south to Canada in the north crossing Texas, Oklahoma, Kansas, Nebraska, South Dakota North Dakota, Colorado, Wyoming and Montana; approximately 3000 kilometres south to north (25° N to 55° N latitude) and 1200 kilometres from east to west (95° W to 105° W longitude). It is an ecological transition from tall grass prairies to the Rocky Mountain range. The southern Great Plains are under the influence of the Gulf of Mexico that affects relative humidity and rainfall over wide areas. Rainfall drops sharply across this region from over 112 centimeters annually in the east to less than 30 centimeters annually in the west.

Crops, production systems, and markets – Although many crops are grown locally over small areas, the vast majority of acreages in the Great Plains are planted to wheat, maize, sorghum, and soybeans. This is an area of intensive agricultural production systems including an array of established crop rotation schemes and cultivation practices. It is common to find very long-term monoculture (e.g., over 50 years of continuous wheat or maize) adjacent to long established rotation patterns (e.g., two, three, or four crop sequences) integrated with different tillage practices (e.g., no tillage, minimum tillage, full cultivation of soil annually). Within the Great Plains Region, plant-based agricultural systems are fully integrated with animal-based agricultural systems that are highly developed and productive. Consequently, the animal production systems are dependent upon plant based agriculture for feed, forage, and grazing. In some local areas of the Great Plains, as much as 50% of the commodities produced are exported to nations all over the world. This includes grains for animal feed, grain for human consumption, as well as seed for crop production in other nations.

Possible and visible threats – Climate and elevation vary dramatically over this region resulting extremely variable weather patterns that impact disease development and spread. The great variation in latitudes from Mexico to Canada, results in sequential planting dates of the major crops going from south to north.

The consequence of this sequential planting along with the predominant south to north winds during the growing season facilitates the spread of pathogens and insect pests from disease outbreaks in the south across the entire Great Plains to Canada. This route of spread was first recorded for rust pathogens of wheat and is termed the "Puccinia Pathway". Recent epidemics of stripe rust of wheat (*Puccinia striiformis*) and southern rust of maize (*Puccinia polysora*) have caused very significant losses in the hundreds of millions of dollars annually. Of recent concern is the new Ug99 race of stem rust of wheat (*Puccinia graminis* f. sp. *tritici*) that emerged in eastern Africa. Most wheat varieties are susceptible to Ug99; if it gets introduced into the southern Great Plains, the potential to spread throughout the region and cause very severe losses is significant.

Preparedness – Large-scale and small-scale disease surveillance systems are present across the Great Plains. The USDA Cereal Disease Laboratory in Minnesota actively scouts and reports on disease outbreaks and spread through each growing season. Researchers and extension personnel at Land Grant universities within the states manage local disease monitoring systems. Communication systems allow for the dissemination of information regarding disease development at least for the major crops; i.e. wheat, maize, sorghum, and soybeans. The Pest Information Platform for Extension and Education (PIPE) is a national disease surveillance system comprised of an internet-based reporting system and a sentinel plot monitoring system (Isard et al. 2006). Originally designed for monitoring the spread of Asian soybean rust, it will expand to include additional diseases and host crops in the future.

Due to the transnational nature of the Great Plains Region, international cooperation is important to the effective management of these pathogens. Canada, Mexico, and the USA have agreements that permit the exchange of information and expertise regarding high consequence plant pathogens and insect pests. Efforts are underway to strengthen these agreements to help manage the increased threats all nations face.

6. The Global Perspective

Area and boundaries – The term coin "global village" is a perfect demonstration of the global perspective, also regarding crop biosecurity. Virtually, the globe is divided to continents separated by oceans, and islands, which are surrounded only by water. Yet, a region is characterized in many cases by open borders and free move of people plants and plant products. In some cases however, hostility among neighboring countries keep close borders along with lack of knowledge transfer, also with regard to pests and pest threats.

Crops, production systems and markets – A region has wide crop diversity and crop production systems. However, there are some principal crops which are important to most countries in the region. For example date palms are an important export crop in North Africa and the Middle East countries. Commercial treaties among countries in the region include trade in fruits and plant products. This involves also propagation material and artificial substrates which can contain invasive pathogens. Seeds are one of the most important trades among countries.

Possible and visible threats – The threats to crop biosecurity from the global perspective include the complete reservoir of pathogens which exist in any place on earth. The majority of new pathogen introduction all over the world is due to human activities, particularly trade transport and travel. Many pathogens are still confined to specific places. The prevention of their spread outside the original habitat is feasible. However many important other pathogens have already spread worldwide. Thus the control of their further spread is almost impossible.

Preparedness – Plant systems are the foundation of food security for every nation (Stack and Fletcher 2007). In poor nations, plants are the primary consumption food staples while in richer nations plants are the feed for the animals that comprise a large part of the diet. In either case, when plant system fail food affordability and/or food availability decline. The extremely large-scale, the vast complexity of plant systems, and the wide diversity of governments and plant protection programs make obvious the challenge in achieving effective and sustainable global plant health systems. However, the global nature of agriculture and the interdependence of nations for sustainable and affordable food production capability dictate cooperation. Raw material inputs (e.g., seed) and finished plant products are imported and exported across most national boundaries on a daily basis, thus ensuring a constant dissemination of pathogens and pests. Response to intentional introductions will necessitate the cooperation of law enforcement agencies from source and target nations.

The very information necessary for preparedness can be used to enable those with the intent to cause damage to a target nation or region. Modern communication technologies have increased access to that same information while modern transportation systems have simplified the ease of dissemination of threatening pathogens. No nation will achieve food security unless at the very least its trading partners are food secure. This will require international cooperation among governments, industries, and scientific societies.

Science will play a major role in attaining global plant biosecurity through relevant research, targeted education programs, and by providing a platform for international cooperation. The 9th International Congress of Plant Pathology to be held in Torino Italy in 2008 is an opportunity to engage scientists from around the world in addressing the global challenge of plant biosecurity.

Below are some illustrations of the global impact of plant pathogens on some important crops.

6.1. WHEAT

Crop importance – Wheat is among the four most important staple food crops globally and in some regions, the most important grain crop. The annual production of wheat grain in 2007 was over 105 million tonnes.

Area and boundaries – Wheat is truly a globally crop. Wheat is produced on six continents under a wide range of environmental conditions over an area of 25 million hectare. The major wheat producers are the US, Australia, Argentina Canada and the EU.

Crop significance, production systems and markets – Wheat production practices vary widely as a function of the environment within which it is grown, intended use of the wheat harvested (e.g., bread, past, grazing), commercial versus subsistence production, and the intended market (e.g., domestic, export). Intensive commercial agricultural production practices include long-term wheat monoculture in heavily tilled fields with significant inputs of fertilizers, herbicides, pesticides, and fungicides. Less intensive and subsistence production systems rely on an array of crop rotation schemes and cultivation practices including, integrated production systems practicing no tillage or minimum tillage with multiple crop rotation patterns (e.g., wheat-soybean-fallow-maize). In many parts of the world, wheat production is integral to food animal production where animals are grazed on wheat in the fall and fed the harvested wheat grain after harvest. In these systems, both the crop and animal systems need protection from disease as the economics of each system is dependent upon the economic stability of the other system.

Possible and visible threats – Wheat's complex genetic structure has impeded the application of modern genetic technology to crop improvement. Wheat genotypes may be diploid, tetraploid or hexaploid and the base genomes may come from several different species. Wheat lags behind maize, soybean, and rice in genome characterization putting wheat at risk to newly emerged pathogens or the redistribution of existing pathogens. Several pathogens of wheat are global in nature including the rusts and powdery mildew. *Puccinia striiformis* (stripe rust) and *Puccinia graminis* f. sp. *tritici* (stem rust) are two pathogens that have caused yield losses on a global scale over the past few decades. On the regional scale, there are many important pathogens that could disrupt global wheat trade leading to local declines in availability of grain for food and feed. Two such pathogens are *Wheat Streak Mosaic Virus* (WSMV) and the *High Plains Virus* (HPV). WSMV and HPV are both mite-vectored and have caused significant epidemics in the Great Plains Region of the USA with very high economic losses. These pathogens are a threat to wherever wheat is cultivated. *Fusarium graminearum* (Fusarium head blight) is a threat from two perspectives as a cause of direct yield loss and as a cause of reduced grain quality through contamination of the grain with mycotoxins that affect both humans and animals.

Preparedness – Local and regional wheat disease surveillance networks are functioning in different parts of the world. These plant disease surveillance networks and internet-based reporting systems serve as effective early warning systems and protect against environmental damage from the over application of pesticides (Roberts et al. 2006). A global communications network that provides access to the surveillance and disease management information would enable each nation to develop preparedness plans appropriate to the prevailing threats. Global cooperation for research in wheat breeding and genomics is essential in developing effective wheat varieties with resistance to the most threatening pathogens and for the many environments within wheat is cultivated.

6.2. POTATO

Crop importance – Potato is the world's fourth most important food crop after wheat, rice and maize because of its high yield potential and high nutritive value. Potato is an economically important staple crop in both developed and developing countries with annual global production of about 300 million tonnes.

Area and boundaries – The origin of potatoes is in the Andes of Latin America and until the early 1990s, most potatoes were grown and consumed in Europe, North America and countries of the former Soviet Union. Since then, there has been a dramatic increase in potato production and demand in Asia, Africa and Latin America. Today, Asia and Europe are the world's major potato producing regions, accounting for more than 80 percent of world production in 2006. Potatoes are grown in about 150 countries throughout the world and over an area of about 19.13 million hectare.

Crop significance, production systems and markets – Potato has a significant impact on the economy and communities. It is well documented that potato blight caused the famine which drove the immigration of huge Irish populations to North America. Being a food staple, trade of potatoes, potato products and potato seed tuber is an important sector in the economy of many countries. It is also an important industry for products such as potato chips and fries. Drastic changes in production and supply in a region may affect many economical and social aspects in the affected area and also beyond.

Possible and visible threats – Over the past two centuries several important viral, bacterial and fungal diseases have migrated with the crop to much production area. Pathogens such as *Phytophthora infestans, Ralstonia solanacearum, Synchytrium endobioticum,* and *Clavibacter michiganensis* subsp. *sepedonicus* have been shown to cause substantial damage to the potato production and globally change the demographic maps. Although these pathogens are now globally widespread, they are still considered exotic in many production areas. Moreover, new races of these pathogens evolve occasionally and can be regarded new pathogens. Thus, the above mentioned potato pathogens pose a visible and immediate threat all global-wise. Other possible threats may rise from the potato origin. Specific potato pathogens are confined to the Andes and are still regarded as exotic. These include *Puccinia pittieriana* (the causal of common potato rust); *Septoria lycopersici* var. *mallagutii* (the causal of potato leaf spot); *Tecaphora solani* (the causal of potato smut). Central and Latin America are an important genetic source for breeding new potato cultivar. However, this can also serve unintentionally for transfer of new pathogens.

Preparedness – Local and regional potato organization and networks are functioning in different parts of the world. The European potato research organization has production and these plant disease surveillance networks and internet-based reporting systems serve as warning systems and detection of new disease and new races of known disease (mainly *P. infestans*). A global communications network provides exchange of information to develop preparedness plans appropriate to the prevailing threats.

6.3. CACAO

Crop importance – Cacao does not represent a crop which belongs to the major food staples nor will its elimination create starvation in any place on the globe. However, the popularity and the global demand for chocolate and the drink cocoa are constantly increasing.

Area and boundaries – Cacao is a tropical, rainforest-inhabiting tree which grows in a limited geographical zone, of approximately 20 degrees to the north and south of the Equator. Its origin is Latin America, however today 70% of the world crop is grown in West Africa (Ivory coast Ghana, Nigeria and Cameroon). Cacao is also grown in Indonesia, Brazil, the Dominican Republic, and Central American countries.

Crop significance, production systems and markets – The Cacao industry has an important impact from the global perspective. Cacao production is an important sector of the economy of Ivory coast Ghana, Nigeria, Cameroon Indonesia and Brazil (the major producers). It is also an important industry of the two leading cocoa processing countries, the Netherlands and USA. Global wise, changes in Cacao production and supply will affect many economical and social aspects mainly in the rich world.

Possible and visible threats – Over the past two decades several important fungal diseases have gained considerable importance and pose a serious threat to the supply of chocolate. A fifth of the cacao beans used to make chocolate are lost to plant diseases every year. Three major diseases (black pod, witches' broom, and frosty pod rot) can cause substantial damage to the cacao industry (Bowers 2001) The first, Black pod disease, caused by a complex of species of *Phytophthora,* appears where cacao is grown, and is especially severe in West and Central Africa. The two other diseases, witches' broom and frosty pod rot are more confined in their distribution. Witches' broom caused by the fungus (*Crinipellis perniciosa*) is currently found all over South America and the Caribbean islands. However, it does not occur in West Africa. Frosty pod rot, caused by *Moniliophthora roreri,* currently is restricted to northwest South America (Ecuador, Peru, and Colombia) and south Central America (Nicaragua, Costa Rica and Panama). However, it is an immediate threat to Brazil, and a long distance by possible threat to major production areas of West Africa.

The host range of witches' broom and frosty pod rot pathogens includes *Theobroma* spp., *Herrania* (monkey's cacao) species. The host range of witches' broom includes also *Solanum* spp. and Annatto (*Bixa orella*). The versatility in host range enhances the threat for the invasion of these two pathogens to new areas of the main production such West Africa.

Preparedness – The chocolate industry is keenly aware of the requirement for an adequate supply of cocoa beans of consistent quality. Today the main efforts are aimed at the development and promotion of sustainable cacao agriculture, defined by The American Cocoa Research Institute (ACRI) as, "...the production practices in which the small acreage farmer increases or maintains productivity at levels that are economically viable, ecologically sound, and culturally acceptable, through the efficient management of resources. The impact of diseases such as black pod, witches' broom, and frosty pod rot will continue to be devastating until progress is made to control these diseases. Short term solutions such as chemical fungicides are often cost-prohibitive, may be damaging to the environment, and leave undesirable residues in the product. Long term solutions such as biotechnological approaches or breeding for disease resistance are time-consuming to develop and are not readily available for immediate application. The cacao tree is cultivated in a relatively primitive way compared to modern row crop agriculture.

References

Anderson PK, Cuningham AA, Patel NG, Morales FJ, Epstein PR, Daszak P. (2004) Emerging infectious disease of plants: pathogen pollution, climate change and agrotechnology drivers. Trends in Ecology and Evolution 19: 535–544

Ashwal T. (2007) Compling "Clean Arava". Ministry of Agriculture, Plant Protection and Inspection Service (PPIS) online: *http://www.ppis.moag.gov.il/PPIS/Files/News/aravaCleaning.htm.* Accessed on 4 January 2008

Bowers JH, Bailey BA, Sanogo S, Lumsden RD. (2001) The Impact of Plant Diseases on World Chocolate Production, APS Feature story: http://www.aps.org/online/feature/exotic. Accessed on 5 January 2008

Brown JKM, Howmeller M. (2002) Aerial dispersal of pathogens on the global and continental scales and its impact on plant disease. Science 297: 537–541

Cook D, Proctor W. (2007) Assessing the threat of exotic plant pests. Ecological Economics 63: 594–604

Damsteegt VD. (1999) New and emerging plant viruses. APS online. http://www.apsnet.org/online/feature/newviruses/top.html. Accessed on 5 January 2008

Desprez-Loustau ML, Robin C, Buee M, Courtecuisse R, Garbaye J, Suffert F, Sache I, Rizzo DM. (2007) The Fungal dimension of biological invasions. Trends in Ecology and Evolution 22: 472–480

European and Mediterranean Plant Protection Organization (2007). EPPO A1 and A2 Lists of pests recommended for regulation as quarantine pests. *http://www.eppo.org/QUARANTINE/quarantine.htm.* Accessed on 5 January 2008

Fletcher J, Stack JP (2007) Agricultural biosecurity: threats and impacts for plant pathogens. pp 86–94. In: Institute of Medicine. 2007. Global infectious disease surveillance and detection: Assessing the challenges – finding the solutions. Workshop summary. Washington, D.C: The National Academies Press. 263 p

Foxell JW. (2001) Current trends in agroterrorism (antilivestock, anticrop, and antisoil bioagricultural terrorism) and their potential impact on food security. Studies in Conflict and Terrorism 24: 107–129

Frischknecht F (2003) The history of biological warfare. EMBO Reports 4: s47–s52

Isard SA, Russo JM, DeWolf ED. (2006) The establishment of a national Pest Information Platform for Extension and Education. Online. Plant Health Progress doi:10.1094/PHP-2006-0915-01-RV

Jay M, Morad M, Bell A. (2003) Biosecurity, a policy dilemma for New Zealand. Land Use Policy 20: 121–129

Madden LV (2001) What are the nonindigenous plant pathogens that threaten US crops and forests? APS Feature story: http://www.aps.org/online/feature/exotic. Accessed on 8 November 2007

Madden LV, Bosch FV. (2002) A population-dynamics approach to assess the threat of plant pathogens as biological weapons against annual crops. BioScience 52: 65–74

Main CE, Keever T, Holmes G, Davis J. (2001) Forecasting long-range transport of downy mildew spores and plant disease epidemics. APSnet Feature, April-May 2001. Online publication. http://apsnet.org/online/feature/forecast/. Accessed on 7 January 2008

Myerson LA, Reaser JK. (2002) Biosecurity: Moving toward a comprehensive approach. BioScience 52: 593–600

Pheloung PC, Williams PA, Halloy SR. (1999) A weed risk assessment model for use as a biosecurity tool evaluating plant introductions. Journal of Environmental Management 57: 239–251

Pimentel D, Zuniga R, Morrison D. (2005) Update of the environmental and economic costs associated with alien invasive sprecies in the United States. Ecological Economics 52: 273–288

Roberts MJ, Schimmelpfennig D, Ashley E, Livingston M. (2006) The value of plant disease early-warning systems: A case study of USDA's soybean rust coordinated framework. Online. Econ. Res. Rep. No. 18. USDA-ERS, Washington, DC

Schaad NW, Shaw JJ, Vidaver A, Leach J, Erlick BJ. (1999) Crop Biosecurity. APSnet Feature, September-October 1999. Online publication. http://apsnet.org/online/feature/BioSecurity/. Accessed on 7 January 2008

Stack JP, Cardwell K, Hammerschmidt R, Byrne J, Loria R, Snover-Clift K, Baldwin W, Wisler G, Beck H, Bostock R, Thomas C, Luke E. (2006) The National Plant Diagnostic Network. Plant Disease 90: 128–136

Stack JP, Fletcher J. (2007) Plant Biosecurity Infrastructure for Disease Surveillance and Diagnostics. pp. 95–106. In: Institute of Medicine. 2007. Global infectious disease surveillance and detection: Assessing the challenges – finding the solutions. Workshop summary. The National Academies Press, Washington D.C.

Stack R. (2000) Return of an Old Problem: Fusarium Head Blight of Small Grains. APS online. http://www.apsnet.org/online/feature/Fusarium head blight /top.html. Accessed on 7 January 2008

Tsror (Lahkim) L, Aharon M, Erlich O. (1999) A survey of bacterial and fungal seedborne diseases in imported and domestic potato seed tubers. Phytoparasitica 27: 215–226

United States Department of Agricuture, Animal and Plant Health Inspection Service (2007) online: http://www.aphis.usda.gov/programs/ag_selectagent/ag_bioterr_toxinslist.html. Accessed on 14th January 2008

Vacher C, Brown S, Hochberg ME. (2005) Avoid, Attack or do both? Behavioral and physiological adaptations in natural enemies faced with novel hosts. BMC Evolutionary Biology 60: 1–10

Wheelis M, Casagrande R, Madden LV. (2002) Biological attack on agriculture: Low tech high impact bioterrorism. BioScience 52: 569–576

Whitby SM. (2001) The potential use of plant pathogens against crops. Microbes and Infection 3: 73–80

Wolken WAM, Tramper J, Van der Werf M. (2003) What spores can do for us. Trends in Biotechnology 21: 338–344

Yang XB, Dowler WM, Royer MH. (1991) Assessing the risk and potential impact for an exotic plant disease. Plant Disease 75: 976–982

THE NEED FOR DIAGNOSTIC TOOLS AND INFRASTRUCTURE

FEDERICO TINIVELLA, MARIA LODOVICA GULLINO
Center of Competence for the innovation in the agro-environmental sector (AGROINNOVA), University of Torino, Via Leonardo da Vinci, 44, 10095 Grugliasco (TO), Italy

JAMES P. STACK
Biosecurity Research Institute, 1041, Pat Roberts Hall Kansas State University, Manhattan, 66506-7600 KS, USA

Abstract: The increasing threats to agricultural and natural plant systems from introduced pests and pathogens have the potential to destabilize ecosystems and jeopardize achieving global food security. Plant–based agriculture is the foundation of food production systems worldwide. To secure our plant systems, every nation needs the diagnostic infrastructure to ensure the early detection and rapid, accurate diagnoses of disease outbreaks. A comprehensive strategy that includes technology development and deployment, networked diagnostic laboratories, and diagnostician training programs is essential. While traditional symptom-based and morphological diagnostics remain important, nucleic acid and protein based technologies have greatly increased the accuracy and reduced the time for positive identifications. Equally important to the technology employed is a cost-effective and statistically sound sampling strategy. Experience is critical component for effective diagnostics. The very wide spectrum of taxonomic groups from which these plant pathogens arise, underscores the need for access to expertise that can be attained through the establishment of secured communications networks.

Keywords: Field detection, serological methods, DNA-based methods, arrays; globalization, climate change, invasive species

1. Introduction

According to the National Invasive Species Council (NISC 2001), pests of agricultural or natural systems, infectious diseases and agents used for bioterrorism which originated outside a certain territory are defined as Invasive Alien Species (IAS). IAS are nonnative organisms which cause, or have the potential to cause, harm to the environment, economy or human health. IAS represent one of the major threats to the agricultural sector and, to a broader extent, to the economy of an entire country.

Globalization, climate change, land use changes, and advances in certain technologies (e.g., rapid transportation, genetic modification of plants), may increase the risk of biological harm from IAS and make prediction about invasiveness more difficult (Meyerson and Reaser 2002). Increased human travel and global trade present additional major mechanisms for the spread of microbial pathogens, pests and plant seeds with the potential to become invasive and cause significant harm wherever they are introduced. The United States Department of Agriculture (USDA) (2001) reported that tourism has become the world's biggest industry constituted by travelers who can become contaminated with pathogens, seeds or harmful pests which may subsequently escape into the environment.

Climate change is becoming a key factor in driving the emergence of infectious diseases (Anderson et al. 2004). As a consequence, preventing and controlling invasions of harmful organisms will become an even greater challenge due to the alterations of CO_2 concentration, increasing temperature and rainfall intensity, that may modify ecosystems and therefore pest and pathogen spread patterns and behaviour. The introduction of nonnative organisms into the environment can also lead to land use changes, e.g., invading trees can transform grassland into forest and thus greatly reduce the supply of surface water available for drinking and irrigation (Dukes 2000). Finally, the genetic modification of plants or microbes could increase the risk of infestations by enhancing an organisms' competitiveness or through hybridization with native species (D'Antonio et al. 2001).

Given all these factors, it is clear that the early detection of new plant diseases (introduced by either accidental or deliberate means), is a basic principle to avoid severe disease outbreaks, and is becoming more and more challenging. An effective early detection system is based on well developed capacities for taxonomic identification of harmful and potentially harmful organisms and on effective monitoring activities which allow the detection of organisms of concern through the evaluation of vulnerable sites or pathways of potential invasion (Meyerson and Reaser 2002). These recommendations are valuable after a previous identification of those pathogens posing the greatest threat to a country's crops. In Europe, within the framework of the EU funded project "Crop-Bioterror", a list of plant pathogens that may pose a real threat to European agriculture and forestry was identified together with a risk assessment methodology specifically adapted to agroterrorism (Latxague et al. 2007).

The diagnostic infrastructure necessary to ensure early detection and accurate diagnosis will be considered starting from diagnosis under field conditions and to the most recent and advanced techniques available.

2. Field Detection

Identification of affected plants in the field is one of the first steps in diagnosing a plant disease. Both scientific and common names of the plant should be noted. In addition it is important to know the specific variety or cultivar, whenever possible, since a great variation in susceptibility to a specific disease may occur within different cultivars and plant species (e.g., all wheat cultivars are not susceptible to all races of *Puccinia graminis* f. sp. *tritici* causing stem rust of wheat). Knowing the identity of the plant species affected allows the pathologist to utilize various sources that contain lists of plant diseases associated with specific plants (e.g., the APS Press Compendia or the CABI Crop Protection Compendium) (Riley et al. 2002).

Diagnosis of plant diseases in plants showing symptoms can be relatively simple when typical, definitive symptoms are evident. However, symptoms are not always unique and can be confused with other diseases, as in the case of halo blight of beans, caused by the regulated *Pseudomonas syringae* pv. *phaseolicola*, and brown spot of beans caused by the unregulated *P. syringae* pv. *syringae*, cited by Schaad et al. (2003). Diseases also involve a progression of symptoms, which can vary significantly and can help in distinguishing the presence of a pathogen or other injuries caused, e.g., by herbicide application.

Attention must be paid to variations in symptoms which can be due to the presence of different factors at the same time, leading to a symptom expression other than the one caused by the diverse pathogens acting separately. The presence of signs of biotic causal agents, such as mycelia or fungal spores for plant pathogenic fungi and pests themselves, insect frass and eggs for insects, is a more important element than the symptoms themselves in the diagnosis of a disease and in the identification of the agent causing the disease. Finally, two more elements have to be considered when identifying plant pathogens or pests: distribution of symptoms over the affected area (patterns) and host specificity (Riley et al. 2002).

3. Morphological Analysis

In the case neither symptoms nor signs provide enough specific or characteristic information to define the cause of an infectious plant disease, it may be necessary to test the sample in laboratory in order to isolate and identify the causal agent. This can be a time-consuming and labour-intensive process which provides specialized skills.

Steps for pathogen isolation and identification through morphological analysis can be summarized as follows:

- placing of sample under conditions that will allow an infectious agent to grow and possibly induce sporulation;

- drawing of pieces of infected plant tissue (possibly from the margins of the diseased tissue) and plating on nutrient and/or selective media;
- isolation of organism growing out of the tissue in pure culture;
- conducting Koch's postulates;
- identification of the pathogen.

Identification of the pathogen is based on the observation of its main morphological characteristics through a microscope with a suitable magnification and then comparing these traits with the reference features which have been previously determined and which can be found in the literature.

The accuracy of morphological identification is attenuated by the requirements posed by invasive species monitoring: the difficulty in identifying early life stages of pests (which is necessary for tracking invasions) or the time demanding procedure for growing plant pathogenic fungi or bacteria on nutrient media are crucial issues which must be carefully considered when time is the key factor (Darling and Blum 2007). These limitations have led to the development and to the use of different tools for monitoring invasive species, that is immunological methods and DNA-based approaches.

4. Serological Methods

Serological tests (or immunoassays) were originally used for viral disease diagnosis before DNA-based techniques were available. They are based on the uniqueness of proteins on the surface of the pathogen and rely on the identification of a pathogen (containing an antigen) through its reaction with specific antibodies. The most common serological test for the diagnosis of plant diseases is the enzyme-linked immunosorbent assay (ELISA), which can use either polyclonal or monoclonal antibodies or a combination of both. ELISA tests use an enzyme linked to an antibody, which produces a colour reaction in the presence of the pathogen. Polyclonal and monoclonal antisera for many viruses have been developed for commercial use and have been used in many protocols to identify virus, including immunodiffusion assays, western blots, dot-blot immunobinding assays, immuno-strip assays and serologically specific electron microscopy (Schaad et al. 2003). Advantages of this kind of test include specificity, speed and sensitivity: relatively low concentrations of antigen can trigger the colour reaction, even though the sensitivity of an ELISA assay varies depending on the organism (e.g., it is 10^5 CFU/ml for bacteria) (Schaad et al. 2003); besides the assay can be automated for large-scale testing, as it is required in programs that certify seed or vegetative plant parts free of certain pathogens (Schumann and D'Arcy 2006). Other formats for immunodiagnostic include immunofluorescence and immunofluorescence colony staining. They use fluorescent-labelled antibodies that react either directly with cell antigens or indirectly with anti-rabbit globuline and permit direct visualization of cells through a fluorescence microscope. Immunofluorescence colony staining is a combination of colony counting in a selective nutrient medium and immunofluorescence staining: the aim is to obtain fluorescent bacterial micro-colonies which can be detected under UV microscope. Immunofluorescence techniques are especially useful for detecting seed-borne and tuber-borne bacteria (Veena and van Vuurde 2002). For a quick pathogen identification (few minutes) immuno-strips can be used; they carry the antisera bound onto a narrow piece of plastic and produce colour in a horizontal band if the pathogen is present. The technique is cheap and requires little labour or knowledge. As a whole serological techniques can greatly reduce the time needed for diagnosis; however, it should be always taken in consideration that both false positives and negatives are possible.

5. DNA-Based Methods

Nucleic acid-based (primarily DNA-based) diagnostic tools have been widely adopted for applications in research labs, regulatory labs, and diagnostic clinics. DNA-based diagnostic have increased the accuracy of pathogen identifications and decreased the time for most diagnoses. These technologies are likely to be further adopted in the near future. A long list of DNA-based technologies appropriate for invasive species monitoring have been reviewed (Darling and Blum 2007) and include: PCR/RFLP (Polymerase Chain Reaction/Restriction Fragment Length Polymorphism), SSP (Species-Specific PRC), SSCP and DGGE (Single-Strand Conformational Polymorphism and Denaturing Gradient Gel Electrophoresis), DNA barcoding, qPCR (quantitative PCR), Shotgun barcoding, T-RFLP (Terminal -Restriction Fragment Length Polymorphism), and POA (Phylogenetic Oligonucleotide microArray). The nucleic acid of most plant viruses is RNA. Diagnostic protocols for these viruses often require a preliminary reverse transcription step to generate a DNA sequence with which to conduct the array of amplification-based and/or restriction enzyme-based methods.

These nucleic acid-based diagnostic tools have many of the same advantages and disadvantages as serologically-based diagnostics tools with respect to cost per diagnosis, false negatives and false positives, and the challenge of isolating the target signal from a complex sample matrix. The challenge of high cost per diagnosis associated with many nucleic acid based technologies is a significant issue for many plant diagnostic clinics processing routine plant samples. The high cost of the instrumentation (e.g., thermocyclers) for nucleic acid-based diagnostics has slowed the adoption of these technologies for field diagnostics. More research and development is needed to make these techniques field friendly.

Relatively few of the thousands of plant pathogen species have had their genome sequenced. This limits the ability to develop species-specific or strain-specific probes to aid in the diagnosis of high consequence pathogens. This also limits the development of forensic protocols to aid law enforcement agencies in cases of deliberate introductions.

6. Advanced and Future Technologies

The development of methods based on the molecular characterization of the pathogen and, specifically, those based on the amplification of specific sequences of nucleic acids has been fostered through the years thanks to the availability of on-line databases and of entire sequenced genomes of different microorganisms. Notwithstanding molecular methodologies are not yet applied extensively for routine diagnosis of plant pathogens due to some technical and economic limitations such as the risks related to sample contamination, the interpretation of results and the reliability of the extraction procedures in terms of quantity and purity (Mumford et al. 2006). Besides PCR amplification based methods in most cases detect only one pathogen at a time, while multiple and quantitative methods would be more suitable in order to identify several pathogens at the same time, to quantify the susceptibility level of the host or to determine the potential severity of the disease (Cacciola e Faedda 2007). Real Time (RT) PCR is still relatively expensive and runs just in specialized laboratory. An improvement is represented by portable termocyclers, which are quite inexpensive, easy and quick to use. They are employed for field diagnosis or for the monitoring of pathogens in critical points of different productive chains with particular regards to quarantine organisms or, more recently, to those pathogens which can be intentionally used as biological weapons. Main issues related to the use of such devices are described by Mumford et al. (2006).

The main technical limitation of RT PCR is related to the difficulty in running multiple analysis because of the limited number of fluorescent dyes available (Mumford et al. 2006). Arrays technology, on the contrary, allows to perform several PCR or RT-PCR at the same time and is going to become the benchmark for routinely run diagnostic techniques. For instance the system "TaqMan low density arrays" (Applied Biosystems) can be programmed for the quantitative determination of up to 48 samples and a new array is designed for the identification of more than 100 species of the genus *Pythium* (Lievens and Thomma 2007). Through arrays it is even possible to differentiate pathogenic strains from non pathogenic ones within a certain fungal species as in the case of *F. oxysporum* f. sp. *cucumerinum* and *F. o.* f. sp. *radicis-cucumerinum* (Lievens and Thomma 2007).

Lab-On-a-Chip (LOC) systems, based on the micronization of elements and their assembling in one single device, allow to carry out all the different steps of a physical or chemical process such as PCR or DNA electrophoresis in sequence. Another diagnostic method based on DNA array is patented with the trade mark "DNA multiscan" and is based on the ibridisation of amplification products of ribosomal DNA ITS regions through oligonucleotidic probes; diagnosis takes 36 hours and allows the identification of more than 80 pathogens (fungi and bacteria) of different crops. Development of a microarray technique based on biochip technology for direct diagnosis of all quarantine pathogens (fungi, bacteria, virus, viroids, nematods) of potato is the goal of DiagChip EU funded project (www.diagchip.com) coordinated by Central Science Laboratory, UK. Nanobiotechnologies find application even in the study of host-pathogen interactions and in the analysis of gene expression, e.g., relevant to the pathosystem rice/*Magnaporthe grisea* (www.mgosdb.org).

Some technologies are still in a development stage: Li et al. (2005) studied a bar code system based on DNA nanostructures called dendrimers marked with fluorescent dyes, that will permit the unambiguous identification of a pathogen. Metal nanoparticles biologically activated with functional molecules are already used for medical diagnosis and could be applied even in the phytopathological sector.

7. Diagnostic Networks

The application of sophisticated technologies to plant diagnostics has greatly improved our ability to rapidly detect high consequence pathogens from complex matrices at very low population densities. The development of these technologies is costly and the use of these technologies in diagnostics can generate high costs per sample. Unlike human medical systems where there is high value in individuals, the value in plant systems is in populations or commodities derived from the populations. Consequently, it is not always possible to employ advanced technology for the diagnosis of plant diseases. This is particularly true in developing nations that lack distributed diagnostics infrastructure.

The very large number of host plant species and the even larger number of pathogens that cause diseases in those plant species makes it impossible for any one diagnostician to be an expert in the diagnosis of all diseases in all hosts. One of the benefits of a plant diagnostic network is providing the ability to access specific expertise when a diagnostician encounters plant diseases or pathogens with which they have no experience. This can be particularly important to developing nations where adequate plant diagnostic technologies and expertise may be lacking (Boa 2007).

Disease outbreaks can generate very large numbers of samples in a very short period of time. Most local plant diagnostic clinics lack the capacity to process a lot of samples quickly. This leads to a host of problems including a decline in sample quality and the unnecessary quarantine of uninfected plants.

It is important to communicate the results of a diagnosis as soon as possible in order to initiate response protocols and implement regulatory requirements. However, the premature release of information can cause unnecessary negative impacts. An important component of a plant diagnostic network is the capability for secure communications.

8. Conclusions

Control and eradication measures against plant diseases, introduced either accidentally or deliberately depend on proper identification of diseases and of the causal agents. Therefore, diagnosis is one of the most important aspects of a plant pathologist training (Riley et al. 2006). Proper infrastructures devoted to diagnosis should be in place all over the world and special attention should be devoted in the future to upgrade those serving poor farmers in developing countries (Boa 2007). Actually, the consequences of an agroterrorist act could be even more devastating in the absence of adequate structures and of trained people.

References

Anderson PK, Cunningham AA, Patel NG, Morales FJ, Epstein PR, Daszak P. (2004) Emerging infectious diseases of plants: pathogen pollution, climate change and agrotechnology drivers. Trends in Ecology and Evolution 19: 535–544

Boa E. (2007) Plant healthcare for poor farmers: an introduction to the work of the global plant clinic. http://www.apsnet.org/online/feature/clinic. Accessed on 20 January 2008

Cacciola SO, Faedda R. (2007) Nanotecnologie e diagnostica fitopatologica. Informatore Fitopatologico – La Difesa delle Piante 57 (11): 12–20

D'Antonio CM, Meyerson LA, Denslow J. (2001) Exotic species and conservation: Research needs. In: Soulè ME, Orians GA (Eds.) Research Priorities for Conservation Biology. Island Press, Washington DC

Darling JA, Blum MJ. (2007) DNA-based methods for monitoring invasive species: a review and prospectus. Biological Invasions 9: 751–765

Dukes JS. (2000) Will the increasing atmospheric CO_2 concentration affect the success of invasive species? In: Mooney HA, Hobbs RJ (Eds.) Invasive Species in a Changing World. Island Press, Washington DC

Latxague E, Sache I, Pinon J, Andrivon D, Suffert F. (2007) A methodology for assessing the risk posed by the deliberate and harmful use of plant pathogens in Europe. EPPO Bulletin 37: 427–435

Li Y, Hong Cu YT, Luo D. (2005) Multiplexed detection of pathogen DNA with DNA-based fluorescence nanobarcodes. Nature Biotechnology 23: 885–889

Lievens B, Thomma BPHJ. (2007) Quantification in Multiplex Format as a Challenging Goal for Plant Pathogen Molecular Diagnostic Assays. Pest Technology 1: 17–26

Meyerson LA, Reaser JK. (2002) Biosecurity: moving toward a comprehensive approach. BioScience 52: 593–600

Mumford R, Boonham N, Tomlinson J, Barker I. (2006) Advances in molecular phytodiagnostics – new solutions for old problems. European Journal of Plant Pathology 116: 1–19

NISC (National Invasive Species Council) (2001) Meeting the invasive species challenge. National Invasive Species Council, Washington DC, 24 May 2002, www.invasivespecies.gov. Accessed on 10 December 2007

Riley MB, Williamson MR, Maloy O. (2002) Plant disease diagnosis. The Plant Health Instructor. doi: 10.1094/PHI-I-2002-1021-01.

Riley MB, Williamson MR, Maloy O. (2006) Plant disease diagnosis. http://www.apsnet.org/online/feature/plantdisease/. Accessed on 21 January 2008

Schaad NW, Frederick RD, Shaw J, Schneider WL, Hickson R, Petrillo MD, Luster DG. (2003) Advances in molecular-based diagnostics in meeting crop biosecurity and phytosanitary issues Annual Review of Phytopathology 41: 305–324

Schumann GL, D'Arcy CJ. (2006) Essential Plant Pathology. The American Phytopathological Society, St. Paul, Minnesota, pp. 309–314

CROP BIOSECURITY: CONTAINMENT AND ERADICATION OF INVASIVE PATHOGENS

ABRAHAM GAMLIEL
Laboratory for Pest Management research, Institute for Agricultural Engineering, ARO Volcani Center, P.O. Box 6, 50250 Bet Dagan, Israel

JACQUELINE FLETCHER
Oklahoma State University, Stillwater, 74078 OK, USA

Abstract: Introduction of an exotic pathogen may pose a threat to a country's agricultural, economy and trade. Therefore, the main goal when new pathogen is detected is its ultimate eradication. Successful eradication of invasive pathogen involves a concerted complex of detection, risk assessment, adoption of the appropriate strategy, and careful execution of all the control procedures. Successful eradication relies on preparedness, i.e. rapid response and the use of the appropriate strategy. Adoption of the strategies to eliminate an invasive pathogen depends on realistic assessment of the effectiveness of the various approaches, and the feasibility for their success. A quantitative assessment of all the factors that influence the eradication process can lead to the adoption of an appropriate eradication approach and strategy. However, other considerations may influence the selection of the most appropriate strategies and measures.

Keywords: Emerging infectious disease, EID, containment, eradication, management, pest control

1. Introduction

Plant disease management refers to reduction of disease progress and limiting its development to below an acceptable economic level. In the case of deliberately introduced pathogens, the ultimate objectives are to contain the pathogen within the invaded site and, consequently, to eliminate it. These objectives are stated in absolute expressions (e.g., "eliminate", or "eradicate") to emphasize a goal of zero inoculum and disease in the invaded area. Although eradication may not be practical, or even possible, we must first aim at this goal, and consider the best approaches to accomplish it. In this process we should consider the dynamics of plant disease, that is, the changes in the incidence and severity of disease in time and space. Furthermore, we need to assess the relative effectiveness of the various approaches and measures for the elimination of a particular pathogen and their various implications on the pathogen cycle and disease dynamics. We need valid means for quantitative assessment of the effects of the control measures, singly and in combination, on the inoculum reduction and progress of disease. Finally, we need to assemble an integrated strategy that considers all the variables of the pathogen's life cycle and distribution. Although failure to achieve eradication may be due to the application of an inappropriate measure, it also may result if the overall management plan did not cover all the relevant factors (Gottwald and Irey 2007). Any individual measure, no matter how perfectly executed, will not provide the final objective if it is not part of a coherent overall strategy.

2. Definition and Approaches

The criteria for designating an effective pathogen were numerically rated as the "Effective Pathogen Index – EPI" by Schaad (1999). This rating scheme considers important survival, establishment and spread parameters that favour maximal damage of the pathogen in the introduced area. Crop biosecurity is the opposite side of the equation, having the goal to rapidly override all these "effective pathogen virtues" and eventually to eradicate the microbes. The approaches to prevent the establishment of an invasive pest in a new area are based first on preventing the invasion, and then exclusion following its introduction. The following terminology further elucidates those approaches.

- *Prevention (also referred to as Exclusion)* – mainly intended to prevent the introduction of inoculum in a pathogen free area. This approach is focused at blocking the possible pathways of new inoculum and carried out by means of quarantine, pest-free seeds and transplants. Thus, any initial inoculum is destroyed before it arrives or established in the new area.
- *Containment* – applied after an invasive pathogen has been introduced into a new area, with the objective to prevent and restrict its distribution from the infected area to new loci. Containment is accomplished by measures to reduce the initial inoculum of the invaded pathogen, and activity to prevent new infection and spread beyond the initially infected site.
- *Eradication* – aimed at eliminating the existing inoculum of an invasive pest from the contained area. This approach is accomplished by, elimination of all the existing, as well as potential, inoculum of the pathogen in

all sites and host within the area and a buffer zone (including possible vectors and other hosts).

- *Management* – the second phase of action following a successful execution of containment and eradication. The main focus is to maintain pest control procedures to complete the eradication and prevent new emergence of the pest inoculum. Alternatively, this strategy is employed when the invading pest has already spread beyond the area of practical eradication. In the latter situation, a pest management approach, as applied for any existing pest, is performed.

Eradication is the preferred goal in any invasion of a new pest, and accomplishing it depends upon preparedness and rapid response. Moreover, other considerations may dominate the decision of the adopted strategy as is further discussed in this chapter.

3. Critical Aspects in the Eradication of Invasive Pathogens

Several factors contribute to an exotic pathogen's "success" in establishing and causing damage. These factors also dictate the type and extent of measures to be taken to contain and eradicate the introduced pathogen. The following aspects should serve as guidelines for evaluating the threat and assessing the appropriate response. Most of these aspects were described in Chapter 4 while referring to the pathogen. In the following section these factors will be argued with relevance to the response measures.

- *Type of the threat.* Four types of threats exist: food safety, trade, and economic damage to product yield quality and loss of consumer confidence. A threat to biodiversity in a forest or other natural habitat may not be considered as critical as a threat from food toxicity. In the latter situation, the response should be swift, as human and animal health are at the top of the priority ladder.
- *The pathogen.* The systematic grouping of pathogens (i.e. viruses, bacteria, fungi) sets not only the type and magnitude of the agricultural threat, but also the type of response. While viruses can be directly controlled, all the measures are directed to eradicate the vectors and destroy the infected crop. In contrast the strategy to eradicate myctoxin-producing fungi involves activities both targeting the fungus directly and managing the contaminated products.
- *Size and location of the infected area.* The success of containment and eradication measures is inversely correlated with the size of the infected

area. When a pathogen is detected in a small and confined area, a rapid response could be successful. However, a wide area of infection, or multiple infection sites, implies that distribution of the pathogen occurred beyond the detected location. In the second scenario the chances for success of containment and eradication are lower.

- Deliberate introduction of a pathogen into an urban area or forest could be much more difficult to handle than one in a field setting, since other factors may dominate the response approach. For example, the fact that the Florida citrus canker outbreak was within an urban area with many back yard citrus trees was one of the main reasons for the eradication failure during 1995–2001 (Gottwald et al. 2001).
- *Pathogen biology and disease epidemiology.* The ultimate objective of containment is to suppress new infectious inoculum. To apply appropriate countermeasures and accomplish this goal requires knowledge of the pathogen and host biology, life cycle and disease progress. Soilborne fungi have a relatively slow, special distribution pattern. Their containment can be accomplished if the inoculum is suppressed. In contrast it is much difficult to contain foliar disease, such as Karnal bunt (*Tilletia indica*), in which large masses of spores are produced.
- *Vectors.* The involvement of vectors in the disease cycle increases the EPI of an invading pathogen, as it favours rapid spread in and beyond the infection site. Therefore, it is critical to prevent the entry of the vector to the infected area, or to eradicate if it is already there. *Xyllela fastidiosa*, the bacterium that causes Pierce's disease of grapevine, does not occur in Europe or the Mediterranean (EPPO 2008). However, numerous species of Cicadellidae and Cercopidae known to be vectors of *X. fastidiosa*, (Hopkins and Purcell 2002) reside in Europe or the Mediterranean. Therefore, vector management should play an important role in any preparedness and eradication program if and when this pathogen invades a new territory.
- *Other hosts.* Pathogens can infect, survive and spread on hosts other than the main crop. The range of pathogen hosts can include cultured or wild plants that are botanically close (or not), and a wide spectrum of weeds. Failure to identify and eradicate all host species from the invaded area can result in failure of the overall eradication process. For example, because *Phytophthora ramorum*, the causal agent of sudden oak death, colonizes at least 97 host species (APHIS 2006), containment and eradication must include all the possible hosts in the eradication.

- *Critical environmental or climatic factors.* Unusual climatic conditions can induce, spread or suppress epidemics. For example, citrus canker was further spread in Florida by hurricanes.
- *Available measures and time of response.* The critical two steps in EID's are pathogen establishment in a new area and spread to other loci. Because preventing these events is time dependent, the success or failure will depend on the measures taken and the rapidity of the response.
- *The time passed from infection to detection.* Early detection and accurate diagnosis are crucial to prevent the establishment and dispersal of introduced pests and pathogens to minimize subsequent impact. Once an invading species becomes established in an area it can be difficult or impossible to eradicate. A good example of effective and quick detection is the case of pathogens in propagation material which are detected before their introduction into the soil. In contrast, symptoms of citrus greening (caused by *Liberobacter* sp.) were expressed in a period of 2.5–3.5 months after leaves emerged from buds on diseased trees (Su and Huang, 1990). Furthermore, detection of citrus greening pathogens in asymptomatic tissue is inconsistent by any known method. Molecular detection assays may be complicated, and results are not always reliable. The incubation period (i.e. the time from infection to disease), and the latent period (the time from infection to production of an infectious propagule) further extends the time from invasion, possibly beyond the threshold timing for effective containment and eradication.

4. List of Available Strategies and Measures

As outlined before, a wide array of factors and variables are involved in the introduction and spread of a new disease. Therefore, to successfully eliminate an already introduced pathogen a concerted series of simultaneous as well as sequential procedures should be planned and executed. The eradication program should be systematic, based on spatial and temporal levels, and precisely performed. Such programs require the adoption of all possible strategies and the cooperation of all the sectors and institutions which play a role in the eradication endeavor. Brief descriptions of the procedures involved in the eradication process, along with the spatial and temporal parameters which accompany them, are listed below.

4.1. PREVENTION (EXCLUSION)

Prevention steps take place away from the infected location. The main objecttives are to prevent further pathogen spread and to protect special locations. Thus, prevention applies to the zones around the infected area or to protected objects such as nurseries and other propagation material fields. Prevention procedures include:

- Quarantine on any items that can transfer the pathogen into the protected area (machinery, equipment, packing materials, plants or plant products such as fruit).
- Segregation of farms or nurseries to prevent infection and minimize pathogen spread. In a greenhouse farm, division of the houses and assigning workers only to certain designated ones will prevent easy movement of pathogens and their spread within the farm.
- Sanitation and disinfection of all tools, equipment, machinery used in the farm and in other fields. This applies specially to heavy rigs, used to cultivate vast areas, that can transfer the inoculum by moving soil particles, infected grains and more.
- Introduction of only pathogen-free propagation material, and additional disinfection of seed and other propagation material by means of chemicals, thermal treatment or a combination of approaches.
- Treatment to eliminate any source of initial infection in the plant hosts. The main procedure should include intensive pest management in and around the area, by fungicide, bactericide, or other means.
- Monitoring vectors (mainly insects) and intensive pest management programs to suppress vector populations prevent initial infection by vector-transmitted pathogens.
- Elimination and destruction of wild plants and weeds that can serve as volunteer hosts.
- Introduction of resistant cultivars to further prevent initial infection.

4.2. CONTAINMENT

Containment steps are taken within the infected area and the close outskirts around it (also defined as a buffer zone). The ultimate objectives of containment are to detain the pest within the infected area and block its egress. The first step in containment is geographical delineation of the infected area and the buffer zone to be treated. The containment procedures in the outlined area should cover the agricultural, rural and the urban sectors, to reach and eliminate all the existing inoculum. The procedures for successful containment include:

- *Quarantine.* The main focus is to block every possible escape of the pathogen from the contained area. Quarantine measures are aimed at restriction of entry and exit of machinery, equipment, farm materials and products. Furthermore, it is essential to clean and disinfect vehicles, machinery, commodities and any products that can potentially carry contaminants. An outstanding example is the spread of *Synchytrium endobioticum*, the causal of potato wart, from infected fields by contaminated automobiles (Gennings et al. 1997). *Erwinia amylovora,* the causal agent of fire blight, is spread by infected fruits that are ready for market (Roberts et al. 1998). These two cases demonstrate that care should be taken to address any possible pathogen exit pathway.
- *Sanitation.* As a supplement to quarantine, sanitation aims to reduce and suppress the spread of the pathogen and to prevent infection in pathogen free locations within the infected zone. Sanitation includes disinfection of equipment, machinery and working tools after leaving one area in the farm and before entering another. In large scale farms this procedure applies to heavy machinery, which can transfer inoculum by moving soil particles, infected grains and more. In greenhouse production the infected area should be restricted, with access limited to specific personnel. The working tools should be disinfected to prevent pathogen transmission. Examples of pathogens that can be transmitted by tools include a wide spectrum of viruses, bacteria and fungi.
- *Physical barriers.* Creating barriers to contain inoculum within the infected area is especially important with soilborne pathogens, which can spread by root to root contact. A practical procedure of trenching to disrupt grafted root systems is used effectively to control the expansion pockets of oak trees infected by the fungus *Ceratocystis fagacearum,* the causal agent of oak wilt (Wilson and Lester 2002).
- *Vector control.* Intensive insect control and monitoring is directed at eliminating any vector that transmits the pathogen, and preventing further infection within and outside the infected area. Vector elimination is important especially with insect transmitted viruses, phytoplasmas and spiroplasmas, and fastidious walled bacteria.
- *Destruction and removal of infected plants.* Infected and diseased plants that could serve as inoculum sources should be eradicated to prevent further infection and possible spread outside the infected area. Therefore, destruction and removal of infected plants is crucial for the success of the

eradication (as it will be discussed later). This procedure should cover all the cultivated plants regarded as potential hosts in the outlined area and not just the infected plants. The containment procedures which precede the destruction and removal infected plants (or plant parts – in trees) include sequential steps:

First, an appropriate pesticide should be applied on the infected plants to reduce the pathogen and vector populations and to prevent inoculum spread during the consecutive activities of plant removal and destruction. This approach is part of USDA-APHIS guidelines for the eradication of citrus greening disease, caused by *Candidatus Liberibacter* sp., indicating that the psyllid vectors *Trioza erytreae* and *Diaphorina citri* should be controlled prior to tree removal to minimize pathogen spread (USDA 2006).

Important procedures for containing root diseases and soilborne pathogens include destruction of the root system by soil fumigation and herbicide application. These procedures can suppress both existing and new pathogen inoculum and are especially important with annual crops as they also will minimize the viable inoculum left in soil after uprooting the infected plants. Destruction and removal of the infected plants is the ultimate action in this process (as discussed in the eradication section).

- *Intensive pesticide application program (perennial crops).* There are cases in which whole plants are not removed for many reasons. Thus continuous efforts are made to suppress the internal inoculum, arrest further development of the epidemic, and prevent further infection and spread. Such approaches may be successful when an infected area is small and spread is limited. However, this approach may be a weak ring in the eradication chain. Intensive management of vineyards to control the vectors of *Xyllela fastidiosa*, the causal agent of Pierce's disease, failed to eliminate the disease in California, possibly because affected grapevines were not destroyed and removed (Hopkins and Purcell 2002).
- *Eradication of weeds.* Elimination and eradication of weeds and other plants that can serve as volunteer hosts may be very helpful in managing disease. For example, because *X. fastidiosa* has a wide host plant range in California vineyards (Wistrom and Purcell 2005), eradication of all the possible host species from the infected area will be crucial for containment.
- *Eradication in and around water reservoirs.* Treatment in and around water reservoirs to avoid transmission of inoculum to and through water is essential. First steps should prevent the approach of the pathogen to a water reservoir. Controlling volunteer hosts and pathogen vectors around the water area is recommended. However, this procedure is not feasible in cases where

the pathogen has already invaded large water reservoirs or in area like forests, in which access is limited.

4.3. ERADICATION

Eradication is a key element in the chain of steps toward ultimate pathogen elimination. Eradication procedures are performed both within the affected area and in the outlying area (buffer zone). Eradication can be achieved only if followed by rigorous containment strategies. In fact, containment and eradication overlap significantly, and the measures described earlier as part of a containment strategy may serve also as initial steps for eradication. The following are additional measures contributing to pathogen eradication.

- *Removal and destruction of plants or plant parts.* Destruction and removal of the infected plants are required to eliminate the major inoculum source. This process is best initiated by applying a pesticide to minimize the pathogen spread during plant removal. Infected plants are usually destroyed by burning (Schubert et al. 2001), although composting can be effective also (Termorshuizen et al. 2003). It is generally preferable to remove and destroy entire plants. In some cases (such as with certain locally contained tree crop diseases) removal of plant parts can be effective; however such activity will not ensure elimination of all the inoculum. Certain pathogens, such as *Candidatus Liberibacter* sp., have a latent period during which recently infected trees show no symptoms (Bove 2006). Therefore, it is important to delineate an area larger than that containing visibly infected trees, and to remove all the plants in the outlined zone. In practice, several surveys are required over the following days and weeks to identify new infections and remove additional plants.
- *Soil disinfestation.* Soil borne pathogens can survive in soil in many forms, including dormant resting structures. Eradication of pathogens from the soil requires, therefore, a robust treatment such as soil disinfestation using highly toxic soil fumigants with non selective activity. Effective soil disinfestation depends upon establishing proper application conditions (e.g., appropriate soil cultivation, moisture levels, etc.). Disinfestation treatments should be effective down to deep soil layers and should be repeated to assure removal of pathogens that may survive the first application. Until 2005, methyl bromide was an effective and recommended soil disinfestation fumigant. However, due to its ozone depletion potential, this chemical is no longer available. Other fumigants (MBTOC 2007) are available, however, their performance is inferior to that of methyl bromide.

A successful eradication program requires additional measures following the removal and destruction of host plants to confirm the eradication and to suppress the re-emergence of inoculum within the treated area. Because viable pathogen inoculum can still exist in the infected area, and could emerge under favorable conditions, it is wise to further target and eradicates it by additional measures.

- Basic principles of quarantine and sanitation must continue within the eradicated area, and a pesticide application program should be maintained to suppress any inoculum left on plant debris. The latter step is critical in tree crops where trees are pruned rather than removed.
- Monitoring for newly symptomatic plants should be performed routinely, and such plants removed and destroyed. In addition, shoots emerging from remaining plant roots must be suppressed, often by repeat applications of soil fumigants or herbicides. It can be effective also to control weeds and other plants that can serve as alternate hosts for the pathogen.
- Finally, continuous weed control should be practiced as many pathogens can survive and propagate on a number of wild host species.

4.4. MANAGEMENT

Management serves as the second principle of action, following the execution of a successful containment and eradication program in the infested area. The management program prevents the emergence of new inoculum by maintaining conditions unsupportive of an epidemic. Management strategies, applied subsequent to eradication, also provide crop protection tools for a disease recovery plan.

An effective management program should maintain all the quarantine practices as listed above under containment. Sanitation should include disinfestation of tools, equipment, and machinery. Measures performed during the eradication process, including removal of infected plants, pesticide applications against the pathogens and/or their possible vectors, and destruction of weeds and wild hosts, should continue. Intensive pesticide applications are most important in tree crops, where trees were not removed during the eradication process. Additionally management includes:

- *Use of pathogen-free propagation material.* Only certified propagation material should be allowed into the area. Such material should be disinfected by chemicals, thermal treatment or a combination of approaches.
- *Soil treatment.* Soil fumigants or herbicides can be used to destroy the host plant root system, eliminating existing pathogen inoculum and suppressing

the formation of new inoculum. It is also recommended to combine cultural practices and suppressive measures such as compost amendments.

- *Cultural practices.* The cropping system should be modified to maintain conditions that do not favour reemergence of the pathogen. To suppress new infections of *Erwinia amylovora* (the causal agent of fire blight of pome fruit trees), recommendations include reducing fertilization to slow the growth rate of the trees, withholding irrigation water, nitrogen fertilizer, and cultivation (Brunner 1994). Similarly, practices that reduce tree wounding and bacterial movement can reduce the risk of infection. Other cultural procedures may include changes in planting dates and the establishment of wind breakers as mechanical barriers to pathogen movement.
- *Use of resistant cultivars.* Planting resistant varieties of crops is particularly effective in suppressing initial infections, promoting healthy crop production.

5. Assessment and Adoption of Appropriate Strategies

Developing a successful strategy against invasive pathogens requires knowledge of all the factors listed above. In addition, it is helpful to understand the impact level of each specific tactic on each disease element (e.g., the initial inoculum, the infection rate or the vector, etc.), and the implications of the strategy on the possible outcome. Failure to adopt a quantitative approach can lead in many cases to establishment of the EID and its spread over a wide area.

In many cases eradication is sought to protect trade, particularly if a pathogen is new to a country or region (is exotic). After the invasion of an exotic pest a rapid assessment of the potential for pathogen spread and disease epidemiology should be made to effectively counteract an epidemic. If certain elements of the disease are not known, eradication plans may be ineffective and the pathogen may become established and distributed over a large area in spite of efforts to prevent it. If eradication is not a reachable goal, its pursuit will only waste resources. To facilitate the identification and execution of an appropriate management approach, a practical plant disease management "manual" should be available as part of each country's preparedness for crop biosecurity. Such a "manual" will include analyses of all factors relevant to the biology and epidemiology of pathogens of high priority in that country, as well as recommended response plans. A crop biosecurity preparedness manual should include the following parameters.

- Evaluation of critical aspects of pathogen and disease features, limitations and available measures of response (as discussed in section 3 above).

- List of strategies available to manage invasive pathogens (as discussed in section 4 above).
- Grouping of pathogens into clusters defined by similar containment and eradication protocols.
- Practical guidelines for specific responses for each pathogen grouping and each high priority pathogen.
- Guidelines for selecting the most appropriate strategy (eradication, containment, or management) based on the situation.

5.1. GROUPING OF INVASIVE PATHOGENS TO CLUSTER OF MANAGEMENT PROTOCOLS

The list of exotic pathogens that pose a threat to a country is long and covers a wide spectrum of organisms from various taxonomic and systematic groups. EPPO, in its A1 and A2 lists, includes more than 100 pathogens. Obviously, it is not practical to prepare a response protocol for each and every exotic pathogen. However, certain pathogens share many similar characteristics relevant to crop biosecurity and preparedness for response.

We propose to group the pathogens into 9 clusters in which the approach, strategies and measures for containment and eradication are similar. Placement into the 9 clusters would be accomplished according to the following criteria: (i) Type of threat (food, trade, or yield); (ii) Organism group (virus, bacteria, fungus, etc); (iii) Location of the threatened area (agricultural fields, forests, rangelands, etc); (iv) Disease cycle; (v) Relevance of vector(s); (vi) Epidemiology model for the disease; (vii) Pathogen survival potential in the invaded area (viii) Available measures for control. A different and unequal weight would be given to each criterion within each cluster. For example, a threat to food safety, as with mycotoxin-producing organisms, dictates a different response than would be appropriate for fungi clustered in a different group. The involvement of an insect vector in disease spread amplifies the pathogen threat level, since measures are required for eradication of both the pathogen and the vector. Therefore, foliar diseases are separated into clusters regardless of the taxonomic placement of the respective pathogens (bacteria or fungi). Forest diseases are clustered within two separate groups, since their environment is very different from that of agricultural fields.

Table 1 shows the nine disease clusters, with representative pathogens from the EPPO A1 and A2 lists.

TABLE 1. Clusters of Invasive Pathogens Based on Appropriate Management Protocols.

Number	Cluster name	Pathogen and crop
1	Viral diseases	*Potato spindle tuber viroid* (PSTV) – Potato *Tomato spotted wilt virus* (TSWV) – Tomato
2	Bacterial diseases with vectors	*Flavescence dorée* – Grapevine *Candidatus Liberobacter africanus* – Citrus *Xyllela fastidiosa* – Grapevine
3	Foliar diseases (mainly bacteria and fungi)	*Magnaporthe grisea/Pyricularia oryzae* – Rice *Phakospora pachyrhizi* – Soybean *Xanthomonas axonopodis pv. citri* – Citrus
4	Bunt and smut fungi	*Tilletia indica* – Wheat
5	Soil borne pathogens	*Fusarium oxysporum* f. sp. *albedinis* – Date palm *Ralstonia (Pseudomonas) solanacearum* races 3 biovar 2 – Potato, geranium *Synchytrium endobioticum* – Potato
6	Soil borne and airborne pathogens	*Clavibacter michiganensis* subsp. *sepedonicus* – Potato *Leptosphaeria maculans* – Oilseed rape *Cochliobolus miyabeanus* – Rice
7	Mycotoxin producing organisms (mainly fungi)	*Gibberella zeae* – Small grains *Aspergillus flavus* – Corn, soybean *Aspergillus ochraceus* – Corn, wheat, barley
8	Forest foliar pathogens	*Cronartium quercuum* f. sp. *fusiforme* – Pine *Endocronartium harknessii* – Pine
9	Forest soilborne pathogens	*Ceratocystis fagacearum* – Oak *Phytophthora ramorum* – Oak

5.2. PRACTICAL GUIDE FOR SPECIFIC RESPONSES TO EACH PATHOGEN GROUPING AND PATHOGEN

Clustering pathogens facilitates establishment of basic guidelines for effective containment and eradication procedures. Each cluster includes pathogens having similar biology, epidemiology and available control measures. Thus, we present a simple template guide for containment and eradication. The guide, which reflects the characteristics of each cluster, contains many possibilities for pathogen behaviour and spread, and recommends appropriate countermeasures based on these characteristics.

A set of procedures used to assemble the practical guide is shown in Table 2. The first of the two columns contains significant disease variables and serves as an inventory list, directing the user to look for the relevance of each variable in his specific situation. The second column suggests an appropriate response.

TABLE 2. Pathogen/Disease Features and General Implications for a Defensive Response.

Variable	Recommended measures
Area	The treated area should cover all infected sites. If the
Size and number of infection sites	area is remote the whole farm should be placed under
Type of area (populated area,	a quarantine that should extend 3–5 km beyond the
isolated and remote, etc)	infected area
	If the infection covers more than 0.5% of the crop area
	the whole block should be destroyed. If less than 0.5%
	remove only those trees within 30 m of an infected
	tree
Pathogen	
Epidemiology	Foliar application of copper pesticides
Disease progress	Pruning and destruction of infected host or tissues
Systemic infection	Burning of removed trees (*plants?*)
establishment in plant tissue	
Survival in soil	
Vector	
Does the pathogen have a vector?	Detection of the vector
Existence vector in infected area	Control of the vector with pesticide
Transmission with machinery	Re-evaluation of the size of appropriate quarantine and
	treatment areas
Hosts	
Involvement of alternate plant hosts	Treatment of all plant hosts in the area
Are these hosts in the infested area?	
Weeds	
Existence of latent hosts (weeds)	Intensive herbicide control of weeds
Are these hosts in the infested area?	Repeated application of herbicides
Water	
Existence of water reservoir	Possible extension of the containment area
Transmission by water	Treatment around the water reservoirs
	Analysis of pathogen occurrence downstream
Environment	
Current conditions	Spray before (*before what?*) to prevent infection
Forecast for environmental	Uproot plants 30 (*units?*) around infected tree
conditions	
Control	
Is an appropriate chemical available?	Emergency registration of pesticides
Existence of limiting factors	Regulations for chemical use in special areas
Are resistant cultivars available?	Adoption of previous experience (other countries)
Documented management strategy	

Thus, the table will identify a possible response based not only on the nature of the pathogen, but also on the specific conditions encountered in the outbreak. The guide is intended to be a decision-making tool. Thus, for example, the nature of the pathogen influences the area that should be treated for pathogen eradication. If a vector is involved, however, the area subjected to eradication procedures should be larger. The delineation of the treated area will be different for foliar pathogens with vectors than for soilborne pathogens. Table 2 also presents important knowledge gaps and recommended actions. For example, if an appropriate pesticide is not listed, a rapid emergency registration of a specific pesticide may be appropriate.

The decision-making tool in Table 2 is a general guide and template for any pathogen. When applied to a specific pathogen, the first step will be to look at the parameters relevant to that pathogen. For example, the size and number of infection sites will dictate the size and shape of the area to be quarantined. In addition, the existence of a vector will dictate a larger quarantine area. Thus, each parameter in the guide can be regarded as independent in the decision making, and at the same time it will be strongly correlated with other relevant variables.

5.3. SELECTION OF THE APPROPRIATE STRATEGY

Eradication is generally the preferred goal following the introduction of a new pathogen, but may not always be feasible. Eradication steps are often very expensive and there is no guarantee of success. The strategy chosen must be feasible and reasonably likely to accomplish the objectives. The major and initial factor in assessing eradication potential is the specific situation in which the pathogen is found. For example, eradication from a single, isolated locus is much more feasible than from a vast area. Specific characteristics of the pathogens (type of organism, vector, epidemiology, etc.) that influence pathogen spread and establishment, regardless of the initial size and location of the introduction, will impact the likelihood of successful eradication. Since the weights of these factors vary among pathogens, clusters and location types, it is useful to make a quantitative assessment of the probability of successful eradication in a given situation. We suggest here a new value of "successful eradication probability" (SEP), calculated from various elements of the pathogen's characteristics and the specific disease situation. SEP (as shown in Table 3) is a cumulative value based on a hierarchy of criteria specific to the relevant event. It can give a weighted assessment for the probability of eradication success, and indirectly suggest an appropriate strategy. An important component of SEP assessment is any documented eradication effort for the relevant pathogen in other locations and situations.

For each variable of the suggested assessment (Table 3) the SEP can receive a positive score (+), indicating a high probability of success, or a negative score (–), indicating a low probability of success. For each variable the SEP score is inversely related to the probability of pathogen spread and establishment in and beyond the infected area. Factors that enhance pathogen spread and establishment are assigned negative SEP scores, while those that enhance the probability for containment and eradication receive positive SEP scores. Suggested SEP

TABLE 3. Criteria for Assessing the Successful Eradication Probability (SEP), and Suggested Strategies.

Section	Variable	SEP score
Grouping	Pathogen grouping	
Infected area	Size and number of sites	
	Vicinity of susceptible plants	
	Location of the outbreak (populated, remote, etc)	
Pathogen	Epidemiology	
	Disease progress	
	Systemic infection	
	Establishment in plant tissue	
	Survival in soil	
Vector	Does the pathogen have a vector?	
	Existence of the vector in infected area	
	Transmission with machinery	
Other hosts	Does the pathogen infect alternate hosts?	
	Are these hosts in the infested area?	
Weeds	Existence of latent hosts (weeds)	
	Are these hosts in the infested area?	
Water	Existence of a water reservoir	
	Transmission by water	
Environment	Current conditions	
	Forecast for future environmental conditions	
Control	Available chemicals	
	Existence of limiting factors	
	Resistant cultivars available	
	Documented management strategy	
Documented efforts	Cases of success (+) Cases of failure (–)	
Total SEP score		

scores for most of the variables in Table 3, especially those related to the pathogen's nature and features, are shown. However, SEPs for specific circumstances (such as site and size of the outbreak site), will be unique for each specific disease occurrence. Finally, the total SEP score is determined using a mathematical formula that considers the relative weights of positive scores vs. negatives scores. SEP values range from 10 (the highest probability for eradication success) to –10 (indicating that eradication will probably fail).

- Pathogen grouping. Foliar diseases and those having vectors can spread rapidly beyond the delineated boundaries. For such pathogens the SEP score for "Pathogen grouping" will be negative. In contrast, soilborne diseases will usually receive positive SEP scores.
- Infected area. This SEP score depends on the specific situation in the infected area. A positive score will be given to a small infection area and a negative score for a large infected area, or for many sites of infection.
- Pathogen type. A negative SEP score will be appropriate for pathogens having polycyclic disease progress, and a positive score for pathogens with monocyclic disease progress.
- Vector, other hosts and weeds. The existence of alternate hosts and vectors that facilitate pathogen dissemination warrants a negative SEP score, while the absence of a vector will result in a positive score.
- Water reservoirs. The existence of a nearby water reservoir and the ability of the pathogen move through water will result in a negative SEP score, while absence will result in a positive score.
- Environment. The prevalence of environmental conditions that favor rapid and area-wide spread of the pathogen (rain, storm dust), will result in a negative score.
- Available control measures. The availability of effective control measures that can be readily and rapidly applied will result in a positive score. The lack of such measures will result in a negative score.

SEP is not a mathematical model. It is a practical tool for a simple and quick assessment of the probability of eradication success, and indirectly suggests the appropriate strategy. It is structured as a simple arithmetic calculation of the entire assessed factors. In order to facilitate the calculations, the weight of each variable in the overall SEP scoring is identical (although in practice there are difference among their influence on disease eruption and spread). Two examples are given below as case-studies to further assess the validity of SEP:

- *Xanthomonas axonopodis* pv. *citri*, the causal of Citrus canker. A negative SEP score will be given for following variable: grouping (bacteria), pathogen, weeds, environment, control and documented efforts. Thus the initial score of –7 is the baseline (based on the pathogen characterization). If we evaluated the eruption of this disease in Florida during 1995, under SEP scoring, we can give additional negative scores for the infected area (which was large and within urban area). Summing the overall scoring (–8, vs. +2), gives a score of SEP = (–6), which indicates a low probability for success. Indeed, in 2006 the UDSA officially declared that eradication of citrus canker is no longer feasible.
- *Fusarium oxysporum* f. sp. *albedinis*, the causal of bayoud disease in date palm. A negative SEP score will be given for following variable: host and weeds. However a positive SEP score will given for the variables which depend on the pathogen characterization (pathogen, environment, water, vector, and documented efforts). Thus the arithmetic sum of SEP score is +3 as the baseline. When we evaluate the SEP score for the eruption in a given area, SEP scoring can either increase if infected area is confined, and measures for eradication are available. Thus, the probability to eradicate such disease if invaded is higher. Indeed, the bayoud disease is confined to a specific area on the globe and did not establish in other date palm production areas.

6. Conclusions

Although eradication of an invasive pathogen involves many uncertainties, it is usually the default strategy. Unfortunately, if the selected eradication measures fail the result will be the establishment of the pathogen in an area. Preparedness within the agricultural community is critical for success in eradication. A robust preparedness plan depends upon having an organized and effective agricultural management infrastructure, reliable and sensitive tests for disease detection and diagnosis, effective eradication measures, and a cadre of well-trained plant health specialists to implement them. International cooperation and collaborations, such as this project has fostered, will be crucial to the establishment of optimal practices for all.

References

Roberts RG, Hale CN, van der Zweff T, Miller CE, Redlin SC. (1998) The potential for spread of *Erwinia amylovora* and fire blight via commercial apple fruit; a critical review and risk assessment. Crop Prot. 17: 19–28

Anderson PK, Cunningham AA, Patel NG, Morales FJ, Epstein PR, Daszak P. (2004) Emerging infectious diseases of plants: pathogen pollution, climate change and agrotechnology drivers. Trends in Ecology and Evolution 19: 535–544

Bove JM. (2006) Huanglongbing: a destructive, newly-emerging, century-old disease of citrus. Journal of Plant Pathology 88: 7–37

Brunner JF. (1994) Integrated pest management in tree fruit crops. Food Rev. Int. 10: 135–157

Cook D, Proctor W (2007) Assessing the threat of exotic plant pests. Ecological Economics 63: 594–604

Duffy B, Scharer HJ, Bunter M, Klay A, Holiger E. (2005) Regulatory measures against *Erwinia amylovora* in Switzerland. OEPP/EPPO bulletin 35: 239–244

European and Mediterranean Plant Protection Organization. (2007) EPPO A1 and A2 Lists of pests recommended for regulation as quarantine pests. *http://www.eppo.org/QUARANTINE/quarantine.htm.* Accessed on 20 January 2008

Foxell JW. (2001) Current trends in agroterrorism (antilivestock, anticrop, and antisoil bio-agricultural terrorism) and their potential impact on food security. Studies in Conflict and Terrorism 24: 107–129

Gottwald TR, Irey M. (2007) Post-hurricane analysis of citrus canker II: Predictive model estimation of disease spread and area potentially impacted by various eradication protocols following catastrophic weather events. Online. APS feature story

Gottwald TR, Graham JH, Schubert TS. (1997) An epidemiological analysis of the spread of citrus canker in urban Miami, Florida, and synergistic interaction with the Asian citrus leaf miner. Fruits 52: 383–390

Gottwald TR, Graham JH, Schubert TS. (2002) Citrus canker: The pathogen and its impact. Online. Plant Health Progress doi:10.1094/PHP-2002-0812-01-RV

Graham JH, Gottwald TR, Cubero J, Achor D. (2004) *Xanthomonas axonopodus* pv. *citri*: Factors affecting successful eradication of citrus canker. Molec. Plant Pathol. 5: 1–15

Hopkins DL, Purcell AH. (2002) *Xylella fastidiosa*: cause of Pierces's disease of grapevine and other emergent diseases. Plant Dis. 82, 1056–1066

Jeger MJ. (2004) Analysis of disease progress as a basis for evaluation disease management practices. Annu. Rev. Phytopathology 42: 61–82

Jennigs D, Hampson MC, Wood SL. (1997) Contamination of automobiles by *Synchytrium endobioticum* in relations to home gardens in Newfoundland. Can. J. Plant Pathol. 19: 52–56

MBTOC. (2007) Assessment Report of the methyl bromide technical Options Committee (MBTOC). 2006 Assessment of the Montreal Protocol on Substances that Deplete the Ozone Layer. UNEP, Nairobi, Kenya

Pheloung PC, Williams PA, Halloy SR. (1999) A weed risk assessment model for use as a biosecurity tool evaluating plant introductions. Journal of Environmental Management 57: 239–251

Schaad NW, Shaw JJ, Vidaver A, Leach J, Erlick BJ. (1999) Crop Biosecurity. APS Feature article. *http://www. Apsnet.org/online/feature/Biosecurity.* Accessed on 22 January 2008

Schubert TS, Rizvl SA, Sun X, Gottwald TR, Graham JH, Dixon WN. (2001) Meeting the challenge of eradicating citrus canker – again. Plant Dis. 85: 340–350

Termorshuizen AJ, Volker D, Blok WJ, Brummeler E ten, Hartog BJ, Janse JD, Knol W, Wenneker M. (2003) Survival of human and plant pathogens during anaerobic mesophilic digestion of vegetable, fruit, and garden waste. European Journal of Soil Biology 39: 165–171

United States Department of Agriculture, Animal and Plant Health Inspection Service 2007. on-line: http://www.aphis.usda.gov/programs/ag_selectagent/ag_bioterr_toxinslist.html. Accessed on 22 January 2008

USDA-APHIS. 2006. New pest response guidelines *Candidatus Liberobacter africanus*, *Candidatus Liberobacter americanus*, Honglongbing, citrus greening disease. Version 1.1

Wheelis M, Casagrande R, Madden LV. (2002) Biological attack on agriculture: Low tech high impact bioterrorism. BioScience 52: 569–576

Whitby SM. (2001) The potential use of plant pathogens against crops. Microbes and Infection 3: 71–80

Wilson AD, Lester DG. (2002) Trench inserts as long-term barriers to root transmission for control of oak wilt. Plant Dis. 86: 1067–1074

Wistrom C, Purcell AH. (2005) The fate of *Xylella fastidiosa* in vineyard weeds and other alternate hosts in California. Plant Dis. 89: 994–999

Yang XB, Dowler WM, Royer MH. (1991) Assessing the risk and potential impact of an exotic plant disease. Plant Dis. 75: 976–982

THE NEED FOR FORENSIC TOOLS IN A BALANCED NATIONAL AGRICULTURAL SECURITY PROGRAM

JACQUELINE FLETCHER
Oklahoma State University, Stillwater, OK, USA

Abstract: In addition to strong capabilities in prevention and preparedness for the possibility of an intentional introduction of a crop pest or pathogen, it is also important to be able to identify the person or group responsible for the event. The process and impacts of attribution of a criminal act not only help to prevent further actions of a similar nature, but can also serve as a deterrent. Forensic microbiology is a new discipline focusing on criminal actions involving micro-organisms, generally pathogens of consequence to humans, animals or plants. It blends elements of diverse traditional disciplines including, among others, forensic sciences, microbiology, human medicine, veterinary science, and plant pathology. Critical features of a microbial forensics capability include rigorous validation, standardization, reliability, and repeatability of technologies and tools, leading to a high level of confidence in the association of a criminally-introduced pathogen with one or more perpetrators.

Keywords: Attribution, biocrime, evidence, fingerprinting, forensics, forensic investigation, law enforcement, perpetrator, science, suspect

1. Introduction

The focus of this NATO project in crop biosecurity has been to enhance the ability of European and Mediterranean Dialogue countries to avert and avoid acts of biocrime directed against national plant resources. Such crime can fall into a number of categories including biowarfare (state-sponsored programs that are generally reasonably well-funded and supported), bioterrorism (efforts of individuals or of smaller idealistic groups wishing to make a statement), economic crime (efforts of one party – a nation, an individual, or a group – to hamper another's production or marketing of a crop or product for reasons of gaining an economic advantage), or inadvertent crime (illegal import of plant cuttings or grafting material that, without the perpetrators' knowledge, carries a pathogen that subsequently escapes into the environment), etc. While each of these actions involves breaking one or more national and/or international laws, they differ in the level of law enforcement response and the mitigating actions that are called into play. Yet, in each case it is important to identify not only the microbial agent involved, but also the perpetrators of the crime and the route of introduction.

2. What is Microbial Forensics?

Microbial forensics is a branch of forensic sciences that focuses on criminal actions involving microorganisms, generally pathogens. The ultimate goal of microbial forensics is to be able to confidently and reliably trace a criminally-introduced pathogen to one or more perpetrators, and to provide evidence for that association that is sufficiently robust and reliable to prosecute those responsible and thereby bringing them to justice (Budowle et al. 2003; Keim 2002; Murch et al. 2003; Salyers 2003). This process is attribution. A strong national capability in microbial forensics should be viewed as a means to assure that justice is served following a biocrime; it also may contribute to reluctance on the part of potential criminals to repeat such criminal actions, and will serve as a deterrent to those who may have considered this approach to call attention to a cause, to disrupt a society, or to gain economic advantage over a competing entity.

The discipline of microbial forensics is relatively new, having been formed and crystallized as a sub-discipline of forensic sciences by law enforcement and academic scientists in the months following the introduction, in 2001, of the anthrax bacterium, *Bacillus anthracis,* into letters sent through the United States' mail (Budowle 2003 a,b,c; Murch 2002). Five individuals died after inhaling anthrax spores, but hundreds more were potentially exposed and underwent antibiotic therapy, and still more feared exposure (Salyers 2003). As many U.S. citizens learned from newscasts and publications, the strain of anthrax used in those letters was traced, using molecular characterization methods, to have originated in a particular laboratory in Ames, IA. However, that strain of *B. anthracis* had been so widely distributed within the nation's scientific community that it was impossible to determine from which laboratory the bacteria had most recently come. The inability to confidently attribute those cases, which had caused several human deaths and many illnesses, highlighted the need for the law enforcement and security communities to develop a new discipline that would bring the disparate fields of forensic science together with those of microbiology and molecular biology. That discipline, microbial forensics, draws upon a number of established disciplines including microbiology, molecular biology, epidemiology, phylogeny and evolutionary science, public health and medicine, agricultural sciences.

As scientists began to consider what was necessary for a robust capability to trace and attribute crimes involving microbial pathogens, a number of critical elements emerged. Forensic investigations must include procedures for sampling and preserving evidence, maintaining strict chain of custody for samples, understanding microbial background and ecological context, knowledge of the suspect organisms and their phylogenetic and evolutionary history, standardized and validated procedures for microbial strain typing and discrimination with knowledge of test reliability, rational and informed interpretation of evidence, and strong assurances of quality assurance throughout the process. Central to the forensic investigation is the concept of sample matching. Does a microbial sample collected at the site of an event match that found in the garage of the home of a suspect? How similar must samples be, to be considered a match? Are available methods for typing strains, which generally rely upon DNA sequencing or fingerprinting, reliable enough to provide the level of confidence required to make conclusions that will stand up in a court of law? Have such tests been validated and standardized with sufficient rigor to be useful? The answers to these questions, in 2002, were not satisfactory.

2.1. THE SUB-DISCIPLINE OF FORENSIC PLANT PATHOLOGY

In 2002, the United States security community called for an examination of the nation's capabilities in microbial forensics. With foresight and the realization that a society's well-being is dependent not only on the preservation of public health, but also the ability to assure its citizenry an abundant and reliable supply of safe and high quality food, the initiative included assessment of capabilities in agricultural forensics. The latter subdiscipline focuses on the introduction of microbial pathogens to livestock and plants, including crops, forests, rangeland, and other plant resources. With respect to plant biosecurity, a group of senior U.S. plant pathologists, combining expertise that covered the breadth of the discipline, collaborated to assess strategies, resources and capabilities in forensic plant pathology (Fletcher et al. 2006). Although this team found almost no previous history of research or consideration that had been targeted specifically towards applications in forensic science, it was clear that the plant pathology community already had many resources to offer the law enforcement and security communities.

The picture that emerged from the study included strong capability in plant pathogen identification and disease diagnostics, centered around a group of plant disease diagnostic laboratories. Generally, one such lab in each state – often associated with the Land Grant University but in other cases with the State Department of Agriculture – had operated independently for many years but were organized in 2002, by the USDA Cooperative State Research, Education and Extension Service (CSREES) into a system of regional networks coordinated, in turn, into a national network. This system, the National Plant Diagnostic Network, now works in concert with the USDA Animal and Plant Health Inspection Service to monitor, diagnose and report plant diseases in this country. These labs would likely serve as the focal point for an investigation necessary to trace the source of a plant disease outbreak that might appear deliberate.

2.2. WHAT TOOLS ARE NEEDED FOR MICROBIAL FORENSICS?

The emerging discipline of microbial forensics combines elements of many different fields including microbial genetics and genomics, biochemistry and molecular biology, microbial phylogeny and evolution, disease epidemiology and pathogen biology, analytical chemistry, microscopy, and many others. Together, these provide the combinations of scientific technology and resources necessary for forensic investigation. Just as important as having powerful technology, however, are the reliability and validity of the tests, and the skill of the forensic investigator in analyzing the data appropriately. Quality assurance and quality control are necessary hallmarks of a robust investigation.

Until recently the concept of plant pathogens being introduced with intent was almost inconceivable, but the question has become important for many reasons. Law enforcement personnel will be informed and called to the scene only if there is reason for suspicion of illegal activity. Therefore, the first question facing responders to an agricultural or natural resource disease outbreak is whether the event is natural or influenced by human actions, and if the latter, whether unintentional or intentional. What features of a plant disease characterize a normal outbreak and what indicators might trigger suspicion? Some important considerations include the pattern of disease in the field, the nature of the symptoms, the presence or absence of vector insects, the age of the plants, the history of weather conditions in the area, the time of year or season, the normal geographical range of the pathogen, and the rate of disease development. More traditional forensic evidence, such as footprints or tire tracks, broken fencing, discarded containers or other physical evidence, also must be collected.

2.3. THE FIELD INVESTIGATION

Arriving at the scene of a suspected criminal incident involving crops, the forensic investigator must not disturb the scene but should first study the site and surrounding area, taking careful notes on the disease distribution, symptoms, and any other features that might be relevant to the investigation. Photographs should supplement, but never replace, the written record. Interviews with farmers, neighbors, crop consultants, extension personnel and others may provide insights not accessible by other means.

As the investigation continues, many decisions must be made. What constitutes evidence in an agricultural crime? What samples should be collected? Should samples be taken from plant leaves, stems, or roots? What about soil, or water from nearby streams or irrigation pipes? Should neighboring fields also be sampled? How many samples are needed? Should they be taken from a region of the field in which the disease is prevalent, or should sampling be done in a random pattern?

Forensic investigators then face issues of proper sample handling. Important considerations include the type and composition of collection bags or bottles; the nature and composition of collection tools such as swabs, wipes, knives or clippers; the type of collection liquid (if any); the temperature at which to store the samples in the field and during shipping; and the type of shipping container used for delivery to the laboratory. Critical to any criminal investigation is strict maintenance of a chain of custody; accurate written records of every aspect of the collection process and of the individuals who, at any time in the process, have custody of the samples. Failure in this step can mean failure downstream during legal proceedings.

At the forensic laboratory, analytical tests appropriate for that particular sample type – whether based on molecular signatures, physical appearance, or other characteristics – are performed according to highly standardized protocols with rigorous regard to stringency and care. Careful record-keeping remains essential throughout the process. However, regardless of the care with which sample collection, handling, and testing have been performed the investigation is dependent upon careful and experienced analysis of the resulting data. All interpretations must be made within the context of the event, the collection history, the background clutter, and other elements of the investigation.

2.4. NEEDS AND GAPS

For plant pathologists, then, the challenge was to assess the tools, knowledge and resources at hand, which were developed for peaceful purposes and natural disease outbreaks, and adapt them to the more stringent requirements of forensic science. We not only need to identify the pathogen to genus and species, but also to be able to finely discriminate among strains of a pathogen that may be very similar. We need to know exactly how confident we can be of the test results. For many plant pathogens, methods of detection and identification have not been optimized, standardized or validated. In many cases, traditional, time consuming methods such as host range studies and use of sets of "differential" cultivars displaying different reactions to different microbial taxa are still being used. The most striking element of our arsenal of tools is the application of DNA typing and strain comparisons. Still, new, more rigorous and reliable analytical methods are needed. Priority should be given for development of technologies applicable to high-priority plant pathogens, such as those on the "Select Agent" list developed by the USDA APHIS, and to multi-plex tests capable of detecting multiple pathogens or multiple strains on a single test run.

We must understand the rates at which threatening pathogens change over time in natural settings and in culture, and how those mutation rates affect the "match" criteria in a forensic investigation. It becomes important to understand the background clutter of normal microbial inhabitants in a crime setting – we cannot recognize novel pathogens if we do not know what microbes are normally present in that environment.

2.5. STEPS FOR THE FUTURE

Typically, security and law enforcement personnel are well trained in emergency management and crime scene investigation. It is not typical, however, for such individuals to have had training or previous experience in agricultural settings. On the other hand, NPDN and APHIS personnel are comfortable working with farmers and ranchers in rural sectors, but may not have the experience to interact optimally effectively with law enforcement representatives who would become involved were a crime to be suspected. Needed are targeted tools and experiences to bring these two groups to common ground. Training exercises that specifically script the involvement of security personnel and require their interaction with Plant Disease Diagnostic Laboratories will provide valuable opportunities for moving towards a shared understanding of the roles and responsibilities of all involved.

Also helpful would be the development of decision tools to assist law enforcement personnel to work appropriately and effectively in agricultural settings. A tool being developed (Fletcher et al. 2007) for this purpose, designated a "decision tree", will organize, prioritize and focus the investigators' energy to gather crime scene evidence that is useful and trustworthy, to preserve that evidence for use both in real time and in the future, and to support the investigation's outcomes with a solid framework of evidence.

3. International Cooperation

Plant disease epidemiologists understand that international borders cannot be closed to the natural or inadvertent movement of pathogens, which can be borne on the winds, carried by streams or ocean currents, or unwittingly moved by international travelers or trade. Often, the diseases of greatest concern to a nation are not yet present in that country but could cause economic hardship were they to arrive. Being prepared for the influx of a new pathogen – by whatever means – requires learning as much as possible about it, yet regulatory restrictions may make it difficult for plant pathologists to conduct research on exotic pathogens. Fortunately, the discipline of plant pathology comprises a global community in which information exchange and cooperative and collaborative research, are common. This NATO project is just one example of the fact that scientists have a strong history of working together across boundaries. This tradition of international cooperation will be important also in applications to forensic plant pathology – attribution of the *intentional* introduction of a plant pathogen across national borders.

4. Need for Education, Training and Cooperation

Forensic plant pathology is a new component of traditional plant pathology in which we will need to adapt and develop the tools of our discipline to the particular demands of the forensic investigation. Very few are trained in both plant pathology and forensic sciences. There is a need for training at several levels. Graduate programs providing specific coursework relevant to both aspects are needed to prepare researchers for future tool development as well as for the appropriate application or adaptation of existing technologies to this purpose. Targeted training programs for plant disease diagnosticians, extension personnel, and law enforcement also are needed. These will need to address not only the scientific aspects of the issue but also the unique roles and responsibilities of various agencies and responders so that actions at the crime scene are seamless and that appropriate follow-up occurs.

5. Conclusions

Forensic plant pathology is a new sub-discipline of the traditional field of plant pathology. While most plant pathologists are focused on the resolution of problems associated with natural outbreaks of plant disease, with the goal of helping farmers and producers to optimize efficiency, the targeted user of the work products is the law enforcement and security communities with the end goal of identifying the source of the event and attributing responsibility to the perpetrator(s) so that they are brought to justice. Success of the development of this sub-discipline will depend critically on regular interaction and communication between plant pathologists and members of the security committee. Law enforcement must communicate the specific needs of the security community to plant pathologists, while the scientists must target their activities to assure that their work is relevant and usable to solve real problems. While the two have much in common, the missions are different.

References

American Phytopathological Society Public Policy Board. (2002) The American Phytopathological Society, first line of defense. APSnet. http://www.apsnet.org. Accessed on 15 January 2008

Budowle B, Burans J, Breeze RG, Wilson MR, Chakraborty R. (2005a) Microbial forensics, pp. 1–26 In: Breeze RG, Budowle B, Schutzer SE (eds.), Microbial Forensics, Elsevier Academic Press, London

Budowle B, Johnson MD, Fraser CM, Leighton TJ, Murch RS, Chakraborty R (2005b) Genetic analysis and attribution of microbial forensics evidence. Crit. Rev. Microbiol. 31: 233–254.

Budowle B, Murch RS, Chakraborty R. (2005c) Microbial forensics: the next forensic challenge. Int. J. Leg. Med. 119: 317–330

Casagrande R. (2000) Biological terrorism targeted at agriculture: The threat to U.S. national security. The Nonproliferation Rev./Fall-Winter: pp. 92–105. http://cns.miis.edu/pubs/npr/vol07/73/73casa.pdf. Accessed on 14 January 2008

Fletcher J, Bender CL, Cobb WT, Gold SE, Ishimaru CE, Luster DG, Melcher UK, Murch RL, Scherm H, Seem RC, Sherwood JL, Sobral B, and Tolin SA, Plant pathogen forensics: Capabilities, needs and recommendations. (2006) Microbiol. Molec. Biol. Rev. 70: 450–471

Grunow R, Finke E-J. (2002) A procedure for differentiating between the intentional release of biological warfare agents and natural outbreaks of disease: its use in analyzing the tularemia outbreak in Kosovo in 1999 and 2000. Clin Microbiol Infect 8: 510–521

Hughes G, Madden L, Gottwald TR. (2004) Strategies of sampling for detection. Phytopathology 94: S137

Madden LV. (2001) What are the nonindigenous plant pathogens that threaten U.S. crops and forests? APSnet Feature Article. www.apsnet.org/online/feature/exotic/. Accessed on 14 January 2008

Madden LV, Hughes G. (1999) Sampling for plant disease incidence. Phytopathology 89: 1088–1103

Madden L, Wheelis M. (2003) The threat of plant pathogens as weapons against U.S. crops. Annu. Rev. Phytopathol. 41:155–176

Murch RS. (2003) Microbial forensics: Building a national capacity to investigate bioterrorism. Biosecur. & Bioterr.: Biodef., Strateg., Pract. & Sci. 1:117–122

Nutter FW Jr. (2004) Post-introduction mapping of new and emerging agricultural pathogens in real-time using GPS and GIS technologies. (Abstr.) Phytopathology 94: S130

Nutter FW Jr, Madden LV. (2002) Plant diseases as a possible consequence of biological attack. In: Bronze MS and Greenfield RA. (eds.) Biodefense: Principles and Pathogens, Horizon Bioscience, Norwich

National Research Council, Committee on Biological Threats to Agricultural Plants and Animals. (2003) Countering agricultural bioterrorism. National Academies Press, Washington D.C.

Stack J, Cardwell K, Hammerschmidt R, Byrne J, Loria R, Snover-Clift K, Baldwin W, Wisler G, Beck H, Bostock R, Thomas C, Luke E. (2006) The National Plant Diagnostic Network. Pl. Dis. 90: 128-136

Wheelis M, Casagrande R, Madden LV. (2002) Biological attack on agriculture: Low-tech, high impact bioterrorism. Bioscience 52: 569–576
Whitby SM. (2001) The potential use of plant pathogens against crops. Microbes Infect. 3:73–80
Whitby SM. (2002) Biological warfare against crops. Palgrave, Basingstoke

THE NEED FOR SECURE COMMUNICATIONS NETWORKS AND GLOBAL CONNECTIVITY

JAMES P. STACK, WILL BALDWIN
Biosecurity Research Institute, 1041 Pat Roberts Hall
Kansas State University, Manhattan, Ks 66506-7600

Abstract: Secure communications systems are essential to an effective and comprehensive crop biosecurity plan. Communications are integral to early warning systems, the implementation of rapid response plans, and law enforcement efforts to apprehend those responsible for intentional introductions of infectious agents. The global nature of food production and distribution systems requires international networks of communications to facilitate prevention and mitigation of disease outbreaks in crop systems. Challenges exist to establishing and sustaining global communications networks for crop biosecurity. Those challenges include the appropriate design and rational deployment of securable communications systems, especially for nations with underdeveloped infrastructure and difficult terrain. Also important to achieving crop biosecurity will be the development of a basis for global connectivity that transcends political boundaries, geography, and the pressures of economic vitality. New technologies with innovative design concepts may provide solutions to only some of these challenges. The challenge of establishing mechanisms for international cooperation that do not compromise global trade needs to be addressed.

Keywords: Agroterrorism, biocrime, biosecurity, communications network, connectivity, cybercrime

1. Introduction

At the highest level of organization, biosecurity is a state of preparedness that incorporates systems for prevention, detection, response and recovery to pest and pathogen introductions and disease outbreaks (Stack and Fletcher 2007). In addition to well trained personnel, biosecurity systems are comprised of soft and hard infrastructure including communications technologies and networks (Myerson and Reaser 2002; Stack et al. 2006). Effective and secure

M. L. Gullino et al. (eds.), *Crop Biosecurity*.

communications systems are essential to achieve preparedness prior to introductions and to implement effective response plans after an introduction. The successful resolution of disease outbreaks is dependent upon 1) the exchange of information among those responsible for response, 2) the dissemination of information to the impacted populations, and 3) the coordination of communications during recovery efforts. Because of the trade-sensitive nature and economic importance of much food system information, premature disclosure of unverified information can cause enormous economic and social damage whether or not an actual outbreak occurred.

Disease surveillance can form the basis of an effective early warning system (Bandyapahdyay et al. 1999; Main et al. 2001). This provides the time required for the potentially impacted areas to prepare for the introduction of a new pest or pathogen, dramatically reducing the ultimate impact (Isard et al. 2006). Disease severity, the total area affected, and the total economic impacts can be significantly reduced (Roberts et al. 2006). Early warning systems are dependent upon effective communications. There is little utility in generating important information if that information cannot be disseminated to those who need it and protected from those who may misuse it.

In recent years, there has been an increased concern regarding the possibility of the intentional introduction of pests and pathogens targeting food production and distribution systems for political or criminal purposes (Casagrande R. 2000; Madden and Wheelis 2003; Nutter and Madden 2002; Schaad et al. 1999; Whitby 2001). Although few documented incidences have occurred relative to other societal targets (e.g., humans, governmental and financial institutions), to ignore this possibility puts at risk the sustained availability of safe and affordable food. The dispersed and unsupervised nature of many food production and distribution systems makes possible the potential for criminal intervention for economic gain (Fletcher and Stack 2007; Kadlec 1998; Wheelis et al. 2002). In order to minimize the impact of such introductions, rapid response is necessary to keep the impacted area as small as possible. Therefore, in addition to physical and biological security features, biosecurity plans protecting food production systems need effective and secure communications capabilities. The global nature of food production systems dictates that these communications systems transcend geopolitical borders and institutional insularity. Consequently, rapid and secure communications among law enforcement agencies, governmental response agencies, and industry from target and source nations are essential to ensure the prevention of additional introductions, the successful management of the outbreak event, and that those responsible for the introduction are apprehended.

There are many challenges to the design and deployment of communications systems capable of meeting these needs. Few nations have fully vertically integrated food production systems. It is common for raw materials to be produced in one nation, shipped for production in a second nation, and then the finished commodities distributed to other nations. Many nations with emerging economies lack the infrastructure to support rapid and secure communications systems. Innovative approaches based upon new technologies may provide solutions to this challenge. Establishing communications systems in developing nations will not only facilitate early warning communications of outbreaks but also contribute to preparedness in these nations by enabling access to international expertise and resources.

A bigger challenge may concern the interplay between international cooperation and competition for commerce. Agribusiness is an important source of information regarding the geographic distribution of pests and pathogens and the emergence of new diseases. Mechanisms to share information that enables preparedness or activates prevention and/or interception strategies need to be developed. Such communications need to be highly secure in order to protect proprietary information upon which commerce depends while ensuring the timeliness necessary for prevention and outbreak management.

2. Communications Infrastructure and Secure Communications

One key component of communications infrastructure is a system that links the field to the diagnostic lab to the regulatory and response agencies. In the United States there are several such systems including the Plant Diagnostic Information System (PDIS) developed at Kansas State University. Currently, this system connects over 60 university and 3 state government diagnostic laboratories across 34 states in a single communications network. PDIS provides the platform for field to laboratory communications, laboratory to laboratory communications, laboratory to extension specialist communications, and record keeping of diagnostic samples and results. PDIS and similar systems store relevant diagnostic data into the National Plant Diagnostic Network database hosted by Purdue University Center for Environmental and Regulatory Information Systems. Storing the diagnostic data in a national repository allows for epidemiological analyses to detect disease trends, correlate disease data with weather parameters, and identify anomalies in disease occurrence (e.g., progression, incidence, severity, and/or spatial/temporal distribution).

Providing for near real-time communication between individuals in the field and specialists in a diagnostic laboratory is significant capability for a communications system. Field personnel can submit diagnostic requests directly with digital images. When a physical sample has been shipped, the laboratory can anticipate the arrival of the sample and track its progress. Field personnel can access the entire history of diagnostic results directly from the system.

Samples are commonly shipped from a source laboratory to laboratories where specialized diagnostic equipment or expertise exist. Facilitating laboratory to laboratory communications ensures that sample information (e.g., collection location, collection date, symptoms, field photographs and diagnosis) accompanies the physical sample as it travels between laboratories.

There are three approaches for developing a multi-laboratory communications and record keeping system, the centralized approach, the distributed approach, and a hybrid approach that is both distributed locally and centralized for data collection. PDIS uses a single hosted system that connects all laboratories. The advantage with this approach is that the server administrators and software engineers are co-located with the centralized system; this is an important economic advantage.With this system design, a single software update is implemented in all laboratories simultaneously.

With the distributed approach the database and applications servers are located at each laboratory or institution. The primary advantage to the distributed approach is that there are fewer single points of failure that could cause the entire system to become unavailable. The other advantage is that each laboratory has its own server administrators and software developers who can modify the local system to meet local requirements and priorities.

Individuals involved in responding to an outbreak must be able to communicate in a manner that ensures the confidentiality, integrity, and availability of information. This can become a significant challenge when responding to outbreaks in rural communities. Scarcer economic resources, long distances and difficult terrain between rural communities can make buried cable, microwave and high throughput terrestrial technologies economically unfeasible solutions. Other technologies, such as satellite and wireless RF, are improving rapidly; however, they suffer from interference, severe weather, low bandwidth, and overall availability.

Communications relating to an outbreak must be kept confidential. Ramifications to trade and markets can be devastating should the communications be compromised by cyber-criminals attempting to gain intelligence or inject data that can affect markets and/or trade. Trade and market fluctuations can result from a mere rumour of an outbreak long before a confirmatory diagnosis becomes available. "The incidents in Estonia should be viewed as a wake-up call. Whether a cybercrime is staged by a competing nation state, a loosely organized cybercrime business or a lone person, information stored on the networks of governments and nationally important organizations should be considered high-value targets," agrees Dr. Eugene Spafford, executive director of the Center for Education and Research in Information Assurance and Security (CERIAS) at Purdue University (McAfee 2007).

Many individuals use portable computers that connect to public networks. These practices incur significant risks of eavesdropping and loss of confidentiality and/or integrity of the user's data. The diligent use of virtual private networks (VPN), intrusion detection systems, local system firewalls, antivirus systems, effective patch management, and user education are all required to protect the endpoints of a secure communications system.

According to NATO analysts, many governments are still unaware of the threats facing them, and some governments are leaving themselves open to attack: "Many government offices don't even know yet that they are leaking information. Ninety-nine percent of cases are probably still not known. Attackers are using Trojan horse (programs that don't replicate but cause damage or compromise computer security) software targeted at specific government offices – because they are custom-written, these Trojans are not amenable to signature detection and they can slip past anti-viral technologies, so this is a big problem. Hackers have dedicated quality assurance capabilities that they run on all of their malware to make sure that their malware doesn't get detected." (McAfee 2007).

Communications must be protected from unauthorized modification and ensure that only authorized individuals modify shared information. Cyber criminals have become unimaginably sophisticated in recent years. There are real risks that cyber-criminals would change diagnostic results, or inject other erroneous data to affect markets and trade.

Communications systems are not useful in a crisis if they are not available when you attempt to use them. High reliability communications infrastructure with many redundancies are required to address risks of system unavailability due to severe weather, natural disasters, power outages, equipment failure, and scheduled maintenance.

Audit logs are essential to make individuals responsible for their actions within the communications system. Auditing and monitoring the action of individuals within the system provides a useful mechanism for the detection of intruders through the automated detection of suspicious behaviour. If a security breach occurs, activity logs allow for the forensic analysis of the security breach and are required to determine if the data in the system was changed without authorization.

3. International Cooperation

World-wide rural communities are becoming more connected. The United States, most European Union governments and private initiatives have implemented funding programs to provide rural broadband services. For example, 2007 EIU e-Readiness Report showed that strong government support allowed Latvia to move up two places and Romania to move up four places in the e-Readiness ranking since 2006. Similarly, local community broadband programs use funds to enable rural towns and villages to establish local community networks.

The highest throughput technologies to-date requires some form of buried cable. While many agricultural rural communities have opportunities for buried cable, socio-economic status, difficult terrain and other factors preclude many agricultural rural communities from access to buried cable high throughput connectivity.

Wireless technologies for computer networks have evolved significantly since the introduction of the IEEE 802.11 standard in 1997. At a low 0.9 MBit/s throughput rate and a small 100 meter outdoor range radius the technology did not promise to be a wide spread solution for rural wireless connectivity. The newest wireless standards are much more promising and include: 1) IEEE 802.11y, to be ratified in June 2008, with a 23MBit/s throughput rate and an estimated 5000 meter outdoor range radius, as well as 2) IEEE 802.16 with a 72MBit/s throughput rate and an estimated 50 kilometres outdoor range radius.

The technological achievements in wireless throughput rates and range are enhanced by other technologies such as the wireless mesh. WI-FI technologies where each transmitter/receiver access point is tethered to a wired connection will be giving way to the new wireless mesh technology. Wireless mesh technologies allow network access point receiver/transmitters to communicate with each other allowing the entire mesh to operate with a minimum of only one transmitter/receiver access point tethered to a wired connection.

Wireless meshes can allow rapid and relatively cheap deployment of communications services to rural areas. In a mesh, the un-tethered network access point receiver/transmitters only need a battery and solar panel to operate, the antennae provides for all communications requirements both to in-range network access point receiver/transmitters (the mesh) and to in-range wireless computers.

References

Bandyopadhyay R, Frederickson DE, McLaren NW, Odvody GN, Ryley MJ. (1997) Ergot: A New Disease Threat to Sorghum in the Americas and Australia. Plant Disease 82: 356–367.

Casagrande R. (2000) Biological terrorism targeted at agriculture: The threat to U.S. national security. The Nonproliferation Rev./Fall-Winter: pp. 92–105. http://cns.miis.edu/pubs/npr/vol07/73/73casa.pdf. Accessed on 4 January 2008

Fletcher J, Stack JP. (2007) Agricultural biosecurity: threats and impacts for plant pathogens. In: Institute of Medicine. 2007. Global infectious disease surveillance and detection: Assessing the challenges – finding the solutions. Workshop summary. Washington, D.C: The National Academies Press

Kadlec Lt. Col. RP. (1998) Biological Weapons for Waging Economic Warfare. Chapter 10 in: Battlefield of the Future, 21st Century Warfare Issues. Schneider BR and Grinter LE (eds.) Air University Press: http://www.airpower.maxwell.af.mil/airchronic les/battle/chp10. html

Isard SA, Russo JM, DeWolf ED. (2006) The establishment of a national Pest Information Platform for Extension and Education. Plant Health Progress. doi:10.1094/PHP-2006-0915-01-RV

McAfee Virtual Criminology Report. 2007. http://www.mcafee.com/us/research/ criminology_report/default.html. Accessed on 5 December 2007

Madden L, Wheelis M. (2003) The threat of plant pathogens as weapons against U.S. crops. Annu. Rev. Phytopathol. 41: 155–176.

Main CE, Keever T, Holmes G, Davis J. (2001) Forecasting long-range transport of downy mildew spores and plant disease epidemics. APSnet Feature, April-May 2001. Online publication. http://apsnet.org/online/feature/forecast/. Accessed on 5 December 2007

Microsoft Security Intelligence Report January through June 2007 http://www.microsoft.com/downloads/details.aspx?FamilyID=4ede2572-1d39-46ea-94c6-4851750a2cb0&DisplayLang=en. Accessed on 5 December 2007

Myerson LA, Reaser JK. (2002) Biosecurity: Moving toward a comprehensive approach. BioScience 52: 593–600

Nutter FW Jr, Madden LV. (2002) Plant diseases as a possible consequence of biological attack. In: Bronze MS and Greenfield RA (eds.), Biodefense: Principles and Pathogens, Horizon Bioscience, Norwich

OFFICIAL IEEE 802.11 WORKING GROUP PROJECT TIMELINES – January 2008. http://grouper.ieee.org/groups/802/11/Reports/802.11_Timelines.htm. Accessed on 4 January 2008

Roberts MJ, Schimmelpfennig D, Ashley E, Livingston M. (2006) The value of plant disease early-warning systems: A case study of USDA's soybean rust coordinated framework. Online. Econ. Res. Rep. No. 18. USDA-ERS, Washington, DC.

Schaad NW, Shaw JJ, Vidaver A, Leach J, Erlick BJ. (1999) Crop Biosecurity. APSnet Feature, September-October 1999. Online publication. http://apsnet.org/online/feature/BioSecurity/

Stack JP, Fletcher J. (2007) Plant Biosecurity Infrastructure for Disease Surveillance and Diagnostics. In: Institute of Medicine 2007. Global infectious disease surveillance and detection: Assessing the challenges – finding the solutions. Workshop summary. The National Academies Press, Washington DC.

Stack J, Cardwell K, Hammerschmidt R, Byrne J, Loria R, Snover-Clift K, Baldwin W, Wisler G, Beck H, Bostock R, Thomas C, Luke E. (2006) The National Plant Diagnostic Network. Pl. Dis. 90: 128–136.

Technology Insights for Rural Connectivity Eric A. Brewer Professor of Computer Science, UC Berkeley Director, Intel Research Berkeley September 2005 http://arnic.info/workshop05/Brewer_RuralConnectivity_Sep05.pdf. Accessed on 4 January 2008

Wheelis M, Casagrande R, Madden LV. (2002) Biological attack on agriculture: Low-tech, high impact bioterrorism. Bioscience 52: 569–576.

Whitby SM. (2001) The potential use of plant pathogens against crops. Microbes Infect. 3: 73–80.

WIMAX AN EFFICIENT TOOL TO BRIDGE THE DIGITAL DIVIDE, Prepared by Guy Cayla, Stephane Cohen and Didier Guigon November 2005 http://www.wimaxforum.org/technology/downloads/WiMAX_to_Bridge_the_Digitaldivide.pdf. Accessed on 4 January 2008

Understanding Wi-Fi and WiMAX as Metro-Access Solutions http://www.intel.com/netcomms/technologies/wimax/304471.pdf. Accessed on 4 January 2008

THE CASE FOR INTERNATIONAL COOPERATION AS A STRATEGY TO ACHIEVE CROP BIOSECURITY

JACQUELINE FLETCHER
Oklahoma State University, Stillwater, 74078 OK, USA

Abstract: The processes of production and distribution of food, feed and fibre to the world's populace are changing dramatically, as agricultural products once unknown outside of a given region are becoming familiar offerings in community markets around the globe. Trade of such commodities is critical also to the economic well-being of most nations. The biosecurity of these products, therefore, is an international concern that can be assured only by multi-national cooperation and vigilance. Sharing of information, technological advances, and research initiatives across national boundaries will facilitate our ability to provide for healthy and safe products for the people of all nations.

Keywords: Cooperation, cross-boundary, exchange, globalization, international, multi-national

1. A Philosophy of Globalization

In centuries past, plants in both natural and cultivated ecosystems were limited primarily to those endemic to an area. Plant pathogens were similarly regional in their range and impacts. The compartmentalization of species was supported by the relatively small range of human travel and the presence of physical barriers such as mountains, oceans and deserts, which served as effective barriers to the free movement of plants and microbes.

The world is a smaller place today. Rapid intercontinental flights and high-speed trains speed travellers from hemisphere to hemisphere and from one side of a continent to the other in hours. Though skies are already crowded, the popularity of air travel is predicted to double once again by the year 2020. Demographics, too, are changing as immigrants from politically unstable, war-ravaged or poverty-plagued nations seek better lives in other countries. Yet, ties to homelands and to the foods of their first culture may spur strong temptation to tuck a small package of seeds or a cutting from a favourite tree into pocket or luggage. International commerce is booming, and the most-exchanged

M. L. Gullino et al. (eds.), *Crop Biosecurity*.

commodities include thousands of agricultural products. Populations of developed nations now expect the ability to purchase fresh fruits and vegetables, flowers and ornamental plants from January to December, and they demand ever more diverse specialty items.

Today, then, we are seeing unprecedented mobility of people, plants and pathogens. Many of the pathogens that pose the greatest threat to a nation's plant resources – crops, forests, rangelands, and other natural ecosystems – are those that are not yet endemic in that area. Should pathogens make their way to a new location, potential plant hosts are less likely to have any resistance. Incoming pathogens may act synergistically with endemic pathogens on certain plant hosts, and may cause novel syndromes as they encounter differing environments during movement.

With such propensity for natural introductions of new pathogens into a given area it would seem logical for nations to prepare for the inevitable influx of exotic plant pathogens, particularly those that are quarantined. However, most nations have strict regulatory requirements for research on exotic pathogens, limiting such activities to one or a few strictly contained laboratories. Participating scientists must submit to rigorous background checks, certain personnel are not eligible to work on – or even to have access to – such pathogens, and steep costs are associated with implementation of required building and site security measures (locked freezers, restricted access work spaces, rigorous pathogen inventories, etc). As a result, few scientists today feel that the benefits of working on exotic pathogens outweigh the disadvantages. In such circumstances, information generated in countries in which the same high-risk pathogens are already present, and not quarantined, becomes critical. In addition, joint research conducted as cross-boundary scientific collaborations may provide answers to practical questions ranging from disease prevention to the development of disease management tools including breeding for crop resistance.

2. Exchange of Technologies

New and emerging technologies being adapted for use in the global community of plant pathology/plant health include a myriad of molecular tests for detection and diagnosis, many of which are being adapted from similar applications in human and veterinary medicine. For example, extremely sensitive "electronic nose" (e-nose) technology, developed initially to detect microbes in foods, has been modified for the detection of environmental threats such as smog and fumigants. Further, e-noses "smell" what a human nose would rather not, and are now used to detect nuisance odours in areas with a burgeoning swine industry. E-nose technology could be adapted easily for use with agricultural threats, and diseases having specific associated odours and/or volatiles, such as the stinking smuts (Karnal bunt and others), could be detected earlier in the field, providing early response options and preventing the wasteful harvest of a crop of poor quality. It also could enable detection of small amounts of infected material in large batches, such as few infected seeds in a seed lot or a small percentage of infected grain in a silo or shipping carton, enhancing the quality of commodities exchanged among nations.

The basic needs for knowledge and technology are common to all nations; what varies is the ability of each nation to develop or acquire, and implement the newest technologies. Duplication of effort in the development, screening, standardization and validation of molecular or serological tests, is inefficient and costly. Exchange of information and application guidelines will be beneficial in terms of efficiency, expense, and timeliness.

In addition to detection and diagnostics, the emerging new discipline of forensic microbiology (including forensic plant pathology) is another area in which international collaboration will be beneficial. Developed initially by the U.S. security community in the months following the contamination of mail with the spores of *Bacillus anthracis*, the agent of anthrax, forensic microbiology has many similarities with the activities carried out by microbiologists in non-criminal settings. Important differences include the necessity for fine discrimination among pathogen strains, and the need for greater levels of standardization, reliability, knowledge of test sensitivity limits and result confidences, and validation (Fletcher Chapter 8 this volume). Here again, interactions within the international community of science will lead to greater breadth of coverage, avoidance of costly duplication, and the synergism of multiple efforts working in cooperation.

3. Vulnerability Assessment

Cooperative interactions and planning among countries in proximity to one another may be very beneficial in assessing vulnerability to pathogens. The sharing of knowledge related to pathogens of common interest will help to focus limited resources on those of greatest priority. In addition, pooling resources among nations can realize savings for each.

4. Exchange of Information on High Consequence Plant Pathogens

4.1. SOYBEAN RUST

Soybean rust, caused by the fungus *Phakospora pachyrhizi,* defoliates plants prematurely, leading to yield losses, fewer seeds per pod, decreased number of filled pods per plant (http://www.usda.gov/soybeanrust/index.shtml). The disease occurred for many years in Asia, more recently moving into South Africa and several countries of South America. Dissemination is usually by spores, which are carried many miles by wind or short distances in the field by splashing rain. U.S. soybean cultivars are highly susceptible, and USDA economists estimated potential annual losses of $640 million to $2.0 billion if the fungus should enter the country (Livingston et al. 2004). International cooperation was critical to the work of U.S. plant protection specialists, who collaborated with Brazilian scientists and carefully studied the disease in that country to learn as much as possible about the pathogen, its impacts, and methods of disease prevention and management. Mathematical models were developed to predict the northward movement of the pathogen and estimate when it might arrive in southern U.S. states, and intensive preparations were initiated, making soybean rust one of the most-anticipated plant disease introductions in U.S. history. Efforts included dissemination of informational materials to stakeholders, development of enhanced disease detection technologies, training exercises for diagnosticians and responders, and establishment of management and recovery guidelines including fungicide stockpiling. When the fungus did appear in U.S. soybean production fields, even earlier than expected in November of 2004, its arrival was attributed to airborne dispersal caused by the high southerly winds of the 2004 autumn hurricane season. Fungicides have been effective in managing the disease, but plant pathologists and breeders around the globe are working to identify sources of resistance.

4.2. WHEAT RUST STRAIN Ug99

Ug99, a new and highly virulent strain of the stem rust fungus, *Puccinia graminis*, was first discovered in Uganda in 1999. For many years a wide spectrum of wheat cultivars resistant to stem rust have been grown commercially. However, Ug99 has evolved to be highly virulent to all these cultivars, and wheat crops everywhere are susceptible to it. For several years the strain spread slowly across east Africa, but in January 2007 spores blew into Yemen, and north into Sudan. It is expected that airborne spores of Ug99 from the affected countries will soon blow into Egypt, Turkey and the Middle East, on to India, and ultimately to the Far East.

Aegilops, the genus most closely related to *Triticum*, comprises 23 species that include diploid, tetraploid, and hexaploid genomes. *Aegilops* species are known to be rich sources of resistance to various pathogens and pests, and many resistance genes have been transferred from this species to wheat. Recently, joint research of American and Israeli scientists (Olivera et al. 2007) demonstrated that *A. sharonensis* (which originated, and appears to be endemic, only in the coastal plain of Israel and in a few locations in southern Lebanon) has a source of resistance to Ug99. Although Ug99 has not arrived yet in either Israel or the U.S., this collaborative effort emphasizes the effectiveness of cross-national research cooperation in finding an effective response to a disease that may threaten wheat production worldwide.

4.3. PIERCE'S DISEASE

Pierce's disease, an economically devastating disease of grapevines caused by the vascular pathogen *Xylella fastidiosa*, is a significant threat to wine-growers, whose crops are valued at more than $4 billion and associated industries at over $45 billion (USDA Animal and Plant Health Inspection Service 2002). To add to the complexity of the threat, the bacterial causal agent is transmitted in nature by plant feeding insects, primarily sharpshooters and leafhoppers (Ba et al. 2006). Consequently, the traditional "disease triangle," which establishes that disease will depend upon the presence of a virulent pathogen, a susceptible host and a supportive environment, must be transformed into a pyramid in which the presence of a suitable vector species is added. If any vector species is already present in a location, the risk resulting from pathogen introduction (by any means) is much higher. In the state of California, in the United States, famous for its wine production, Pierce's disease had been present for years but caused little concern because the blue-green sharpshooter, the insect vector present in the state, was small, weak and non-aggressive. It was not until the recent introduction into the state of a different vector species, the glassy-winged sharpshooter, a large, long-distance movement, aggressive grapevine feeder, that PD became a high-consequence disease in this area.

5. Building a Network

Nations interact at many different levels. Officials of governments generally communicate in formalized routes, and, because they represent their governments, they provide official viewpoints and approved positions. The formality of such communication is necessary because of the authority and impact of those interactions. Scientist exchanges are typically much less formal. Plant pathologists meet one another at scientific congresses, where focus is on research of common interest and enthusiasm, or they meet while studying for advanced degrees, or serving as postdoctoral researchers, in other countries. Scientists communicate quite differently than do officials, because they interact on a day-to-day basis, in settings that emphasize commonalities of interest. Scientists often make outstanding national ambassadors because they meet one another as peers and share mutual respect for each other and for their science.

6. The Role of International Scientific Societies

One of the most productive means of creating linkages among plant pathologists are the many national, regional and international societies of plant pathology; those belonging to the International Society for Plant Pathology are shown in Table 11.1

TABLE 11.1 Societies of plant pathology affiliated with the International Society for Plant Pathology (Source: ISPP website).

Countries and Associated Societies:	
Argentina	Argentine Crop Protection Society
Australasia	Australasian Plant Pathology Society
Bangladesh	Bangladesh Phytopathogical Society
Belgium	Association pour les Etudes et Recherches de Zoologie et de Phytopathologie
Brazil	Brazilian Society of Plant Pathology
Canada	Canadian Phytopathological Society
Chile	Sociedad Chilena de Fitopatologia
China	Chinese Society for Plant Pathology China Taiwan Phytopathological Society
Colombia	Asociación Colombiana de Fitopatología y Ciencias Afines
Denmark	Danish Society for Plant Diseases and Pests
Egypt	Egyptian Phytopathological Society
Ethiopia	Ethiopian Society of Plant Pathology
Finland	Pant Protection Society of Finland
France	Société Française de Phytopahologie
Germany	German Phytomedical Society
Greece	Hellenic Phytopathological Society
Hungary	Hungarian Society of Plant Protection
India	Indian Phytopathological Society
Indonesia	Indonesian Phytopathological Society
Iran	Iranian Phytopathological Society
Ireland	Society of Irish Plant Pathologists
Israel	Israeli Phytopathological Society

(Continued)

Italy	Italian Phytopathological Society
	Italian Association for Crop Protection
Japan	The Phytopathological Society of Japan
Kenya	Plant Pathology Society of Kenya
Korea	Korean Society of Plant Pathology
Kyrgyzstan	Kyrgyz Society for Plant Pathology
Malaysia	Malaysian Phytopathological Union
Mexico	Society Mexicana Fitopatologia
Morocco	Association National pour la Production, la Protection, et l'Amelioration Vegetale
Netherlands	Royal Netherlands Society of Plant Pathology
Nigeria	Nigerian Society of Plant Protection
Norway	Norwegian Society of Plant Pathology
Pakistan	Pakistan Phytopathological Society
Peru	Asociacion Peruana de Fitopatologia
Philippines	Philippine Phytopathological Society
Poland	Polish Phytopathological Society
Portugal	Portuguese Phytopathological Society
Russia	Russian Phytopathological Society
South Africa	Southern African Society for Plant Pathology
Spain	Society Espanola de Fitopatologia
Sudan	Sudanese Society of Plant Pathology
Sweden	Swedish Society of Biopathology
Switzerland	Schweizerische Gesellschaft für Phytomedizin
Thailand	Thai Phytopathological Society
UK	British Society for Plant Pathology British Mycological Society Association of Applied Biologists
USA	American Phytopathological Society
Venezuela	Sociedad Venezolana de Fitopatologia
Yugoslavia	Yugoslavian Phytopathological Society
Regional/International Societies:	
	American Phytopathological Society
	Arab Society for Plant Protection
	Asian Association of Societies for Plant Pathology
	Asociacion Latinoamericana de Fitopatologia
	Australasian Plant Pathology Society
	International Committee on the Taxonomy of Viruses
	International Society for Plant Microbe Interactions
	Mediterranean Phytopathological Union
	Quebec Society for the Protection of Plants
	Society for Plant Protection in the Caribbean
	Southern African Society for Plant Pathology

These organizations, which vary significantly in membership size, activities, publications, and communications, are a focal point for the congregation of scientists of like interests and commitments. With the ability to share information instantly all over the world these professional groups take advantage of informational websites, interactive online chat rooms, linked home pages, teaching aids, directories, image libraries, and innovative e-publications. The usefulness and applications of these resources are as numerous as our creativity and resourcefulness allow.

Indeed, as economics of scale have resulted in the downsizing and/or mergers of many traditional academic Departments of Plant Pathology with other disciplines, and as faculty members joining such departments are becoming broader in their interests and less focused on plant pathology, scientific societies are becoming more important as the gathering point and the facilitator of the disciplinary community. In the coming years that community will continue to become more global in its reach and more international in its scope.

7. NATO Project on Crop Bioterrorism – a Model for International Cooperation

Many types of international collaboration are both possible and productive. Variability can occur in the number of countries participating, whether those countries share vulnerabilities or differ significantly, the degree to which the participating nations are balanced in technological capability and scientist training/experience, etc. For example, the countries of a single region (such as the nations of the European Union) might band together because of common conditions and priorities. In a different sort of association, two or more distant countries or regions (such as the United States and the EU) might come together to create mechanisms for exchange of information on diseases of differing concern to the participants. Both types are beneficial, but for different reasons.

One of the most effective ways to achieve international scientific cooperation is to provide opportunities and incentives for scientists from many nations to interact and work together in an environment that is comfortable, supportive and respectful of all. Two examples of this concept, both part of the NATO Crop Biosecurity Project, are described below.

7.1. INTERNATIONAL WORKSHOPS

One example of an event that fostered new and supportive international linkages was a Crop Biosecurity Workshop, held as part of the multi-national project on Crop Biosecurity, in Szentendre, Hungary, in April 2007. A major factor in the success of this event was the involvement of the Regional Environmental Cooperative (REC), an international centre located near Budapest. The REC and its capable staff have an established history of bringing representtation of the diverse nations of Eastern Europe together in a team approach to solving problems of an environmental nature in the region. The REC has a strong reputation for stressing balanced views, respect for all nations, and facilitation of meetings of many nationalities. In their facilitation of the workshop on Crop Biosecurity, the REC was able to offer financial support for the participation of both young and established plant pathologists from each nation, permitting attendance by many who could not otherwise have enrolled in the workshop. The goals of this event were three-fold: (1) to allow plant disease diagnosticians from the diverse nations of the region to meet each other and compare national needs in a supportive setting, (2) to allow each nation to present their country's current agricultural situation, and (3) to facilitate the development of cooperation and communication among plant pathologists from Eastern Europe, Western Europe, the United States and the Middle East. By the end of the short conference many of the diagnosticians were chatting easily with one another and with the workshop organizers. There was a new feeling of shared interests and common goals, and a sense of the benefit of working together toward those goals. The success of this event provides a glimpse of the potential cooperation that such interactions, in a comfortable and respectful setting, can spark.

7.2. SCIENTIST EXCHANGES

The NATO Crop Biosecurity project provided opportunities for the principle scientists from each country to visit the other countries of the project, in most cases interacting significantly with other scientists of the host's institution and visiting crop production areas to learn about the agricultural systems of that nation. In addition, a workshop in each nation allowed the project scientists to share the concepts of crop vulnerability, preparedness, and recovery with an audience of in-country plant pathologists and governmental agents. Even more significant, however, was the opportunity for an early-career plant pathologist to spend six months working in a plant diagnostic laboratory in another country. The good will and cross-boundary information sharing that occurred during these events will have long term benefits and serve as springboards for future cooperation.

8. Achieving Food Security for All

Our NATO Crop Biosecurity project, though limited in scope and time, has accomplished important goals. It has opened conversations among scientists and representatives of multi-national agencies, brought focus to the need for cross-boundary interactions and networking for protecting our crops, and set a precedent for future cooperation at the level of scientists and policy makers.

Nobel Peace Prize recipient Norman Borlaug, in 1970, stated that food is the cornerstone of national security. His wise remark of almost 40 years ago still accurately captures the complex, multi-faceted, economy-linked, internationally based nature of our global agricultural enterprise. It is the responsibility of plant scientists to come together within our global disciplinary community to create new systems, creative linkages, and shared resources to protect our most important resource.

References

Ba A, Dhawan N, Esser T, Freeze K, Hanes C, Hardy S, Hill B, Lorrick T, Mochel M, Morgan D, Morris G, Morrow K, Oswalt S, Spencer R, Van Rein J, Veling S, Wynn B, Yacoub R (2006) Pierce's disease control program. Annual Report to the Legislature. CA. Dept. Food and Agric. http://www.cdfa.ca.gov/pdcp/Documents/2006%20PDCP%20Report%20to%20Legislature.pdf Accessed on 15 December

Dorrance AE, Lipps PE, Mills D, Vega-Sánchez M Soybean Rust. Ohio State University Extension Fact Sheet AC-0048-04

International Society for Plant Pathology. Associated societies of the ISPP. http://www.isppweb.org/about_associated.asp. Accessed on 15 December 2007

Livingston M, Johansson R, Daberkow S, Roberts M, Ash M, Breneman V. (2004) Economic and policy implications of wind-borne entry of Asian soybean rust into the United States. USDA Economic Research Service Electr. Outlook Rep. OCS-04-D-02, www.ers.usda.gov

Olivera PD, Kolmer JA, Anikster Y, Steffenson BJ. (2007) Resistance of Sharon goatgrass (*Aegilops sharonensis*) to fungal diseases of wheat. Plant Dis. 91: 942–950

United States Department of Agriculture. (2008) Soybean Rust Site. http://www.usda.gov/soybeanrust/index.shtml. Accessed on 17 December 2007

United States Department of Agriculture Animal and Plant Health Inspection Service. (2002) Glassy winged sharpshooter and Pierce's disease in California. PPQ Factsheet, December 2002. http://www.aphis.usda.gov/publications/plant_health/content/printable_version/fs_phglassy.pdf. Accessed on 17 December 2007

CROP AND FOOD BIOSECURITY RESEARCH: LUXURY OR NEED?

MARIA LODOVICA GULLINO
Center of Competence for the innovation in the agro-environmental sector (AGROINNOVA), University of Torino, Via Leonardo da Vinci,44, 10095 Grugliasco (TO), Italy

ABRAHAM GAMLIEL
Laboratory for Pest Management research, Institute for Agricultural Engineering,. ARO Volcani Center, P.O. Box 6, 50250 Bet Dagan, Israel

JACQUELINE FLETCHER
Oklahoma State University, Stillwater, 74078 OK, USA

JAMES P. STACK
Biosecurity Research Institute, 1041,Pat Roberts Hall Kansas State University, Manhattan, 66506-7600 KS, USA

Abstract: The political interest in security issues has resulted in a shift of research priorities to include also in agricultural research in order to increase awareness of the impacts of plant pathogens and the diseases they cause. Security-related issues are being financed particularly in the U.S., but also the European Union is considering security a high priority within its VII Framework Programme. Since the development of crop biosecurity as a field of research is relatively new, an ethical examination of potential conflicts that may arise in this area of research is appropriate. Therefore an open international dialogue may contribute to the development of collaborative research programmes and of supporting infrastructures for crop biosecurity and may offer the tools needed to defend the agricultural sector against crop bioterrorism.

Keywords: Funding, research, terrorism prevention, international cooperation, infrastructures

Recent international tensions in the context of history have drawn attention to the issues of crop biosecurity and the prevention of biological and agricultural terrorism targeting food supply systems. These issues are complex and require the attention of policy makers and scientists who develop and apply the dual-use technologies that can be used to protect or harm food production systems. Research efforts in the United States of America, in Europe, and in other areas of the world, have begun addressing scientifically the topic of crop and food biosecurity.

Throughout human history, plant diseases have caused large-scale social upheavals reflected in famine and the mass migration of people (Schumann 1991). Knowledge of the risks that plant pathogens pose was based on natural occurrences of disease and accidental introductions of pathogens to geographic areas. Although some governments were aware of the development of plant pathogens as weapons to attack the food supply of other nations, that concept was not widely known or discussed by policy makers or scientific societies. However, in the late 1990's the risk of the intentional use of plant pathogens as biowarfare and bioterror agents became more generally understood and incorporated into policy and scientific programs. Recently the high political interest in security issues has resulted in a shift of research priorities to include not only in health related sectors, but also in agricultural research, thereby increasing awareness of the impacts of plant pathogens and the diseases they cause. The shift in human health research priorities has been received with varying responses from different sectors of the scientific community. In the microbiological/medical community, over 750 scientists sent a signed petition to the US National Institutes of Health (NIH) Director charging that NIH biodefense priorities have led to a crisis for conventional microbiological research. The letter argued that research funds from projects of high public health importance were diverted to biodefense projects, at the expense of high priority basic research. Although present, this concern is less widespread in the agricultural science community because the security programs provide access to new funding sources not normally available to agricultural scientists.

The political trend for financing security-related issues continues to favour research in the field of crop biosecurity, particularly in the U.S. The European Union is also considering security a high priority within its VII Framework Programme. This emphasis represents a significant shift in research priorities from the past. Plant pathology research by military scientists was an integral part of state sponsored offensive bioweapons research programs until the application of the biological and toxin weapon convention in 1972, and probably even after that. Current research for terrorism prevention purposes is strongly centred in both government and, importantly, academic laboratories. The general principle behind security in the agricultural sector is that everyone should be ready to respond to an outbreak. This approach, called preparedness, requires that academic scientists be funded for projects addressing novel detection strategies for rapid identification of threats to agricultural systems, pathogen spread modelling, simulation of disasters, information infrastructures for rapid communication among experts for the identification of dangerous species, and capabilities in forensic plant pathology. The overriding benefit of supporting this research is that the capabilities developed for preparedness for bioterrorism and biocrime (e.g., pathogen detection, disease management, pathogen containment or eradication) also provide the technologies and knowledge necessary for preparedness to manage all emergency situations involving newly introduced pathogens.

This security focus for agricultural research can pose challenges to scientists. Because the development of crop biosecurity as a field of research is relatively new, an ethical examination of potential conflicts that may arise (e.g., the dual use nature of the knowledge and technology developed) in this area of research is appropriate. An open international dialogue may contribute to the development of collaborative research programmes and the supporting infrastructures for crop biosecurity.

Often the highest level of damage and greatest social impact caused by a new pathogen is reported in developing countries where less technical knowledge is available and in places where agriculture is the major source of income for the majority of population. In such countries the main national production and economy is based on agriculture. International consortia are important also for ethical reasons; to avoid increasing the technological divide between rich and poor countries, future food security research should include comprehensive analyses of global food systems and the development and deployment of technologies useful also for the poorest countries which often are less well equipped to manage food emergencies due to plant diseases.

In the past, scientists have played important roles in developing biological weapons, including those targeting plant-based and animal-based food production systems. As a professional responsibility and commitment, scientists must now conduct the research necessary to deter and defend the fundamental basis of human source for food – agriculture – against crop bioterrorism. To make this important contribution to global food security, scientists need the support of policy makers who set the priorities for funding programs. Improved dialogue among scientists and policy makers is needed to raise awareness of the importance of pant-based agricultural systems to food security, the potential threats to crop systems, and the necessary knowledge and technologies required for preparedness. We all will suffer the consequences of attacks on our food production systems; international cooperation is essential. We are all in this together.

Reference

Schumann GL. (1991) Plant Diseases: Their biology and social impact. APS Press, St. Paul, MN

EVALUATION OF CRITICAL ISSUES AND IDENTIFICATION OF RESEARCH AND INVESTMENTS NEEDS IN THE FIELD OF CROP BIOSECURITY IN EUROPE

MARIA LODOVICA GULLINO
Center of Competence for the innovation in the agro-environmental sector (AGROINNOVA), University of Torino, Via Leonardo da Vinci, 44, 10095 Grugliasco (TO), Italy

ABRAHAM GAMLIEL
Laboratory for Pest Management research, Institute for Agricultural Engineering,. ARO Volcani Center, P.O. Box 6, 50250 Bet Dagan, Israel

JACQUELINE FLETCHER
Oklahoma State University, Stillwater, 74078 OK, USA

Abstract: Current European capabilities to detect and respond to agroterrorism are modest with a lack of coordination. The main objectives of research networks set up in the frame of FP6 of the European Union addressing issues related to bio-weapons, agro-terrorism and biosecurity are presented. These programmes can help European Union building the capability to increase preparedness investments in infrastructures, development of national diagnostic laboratories and their networking, and investments in the basic and applied research are needed also to support the policies. Groups of researchers having expertise and interest in crop and food biosecurity should be identified and the networks already existing should be integrated in order to build, at the European level, a solid network capable to deal with the basic and applied aspects of research as well as with communication and training.

Keywords: Projects, networks, Framework Program, infrastructures, funding, European Union

M. L. Gullino et al. (eds.), *Crop Biosecurity*.

1. Introduction

The general susceptibility of the agriculture and food industries to bioterrorism is difficult to address in a systematic way due to the geographically dispersed, yet industrially concentrated nature of their infrastructure, and the inherent biology of growing plants and raising animals. Following an attack, the agricultural sector would suffer economically from plant and animal health losses, and the supply of food and fibre may be reduced. The demand for foods targeted in an attack may decline (e.g., dairy, beef, pork, poultry, grains, fruit, or vegetables), while demand for substitute foods may rise. Economic losses would accrue to individuals, businesses, and governments through costs to contain and eradicate the disease, and to dispose of contaminated products. More losses would accumulate as the supply chain is disrupted from farm to fork. Domestic markets for food may drop, and trade restrictions could be imposed on exports. The economic impact would range from farmers to input suppliers, food processors, transportation, retailers, and food service providers. Current European capabilities to detect and respond to agroterrorism are very modest and spread among many organizations, with a lack of coordination.

2. Recent and Current European Projects Related to Biosecurity

Some networks, supported by EU in the frame of FP6 "policies-issues related to civil protection (including biosecurity and protection against risk arising from terrorist attacks) and crisis management", addressed issues related to bioweapons, agroterrorism and biosecurity, covering different aspects. A project titled Crop and Food Biosecurity defined the first list of high-threat pathogens for Europe as well as a list of crops at risk. Its results serve as a very useful starting point: a selected list of plant pathogens, ranked by risk, for which special security and reporting measures might be required, is now available. The results obtained through this project clearly indicate that Europe must be prepared for the threat of agroterrorism, since economic losses from an agroterroristic incident could be large and widespread.

A project designated "BIOSAFE" (BIOlogical agents: Strengthening the Adequate response to deliberate releases by the establishment of a Framework European-wide) aims to enhance the capability of public health and civil protection authorities to respond adequately to deliberate releases of biological agents by terrorists or states. In pursuit of this aim BIOSAFE brings together, interpret and analyse existing expertise on virulence factors of those pathogens and toxins that may be used in acts of bioterrorism, by means of the establishment of a European-wide preparation and response network and associated database information system. This database will be complemented by information on the disinfectants or treatment options for these biologicals and on drugs, antisera and vaccines to stop terrorist induced infection outbreaks. The ASSRBCVUL (Assessment of the vulnerabilities of modern societies to terrorist acts employing radiological, biological or chemical agents with the view to assist in developing preventive and suppressive crisis management) project was designed to assess the vulnerabilities of the EU to the intended release of radiological, biological and/or chemical (RBC) agents by terrorists or bio-warfare participants. The objective of Support Action BIOSAFETY EUROPE (Coordination, harmonisation and exchange of biosecurity practices within a pan European network) was to provide data for the harmonisation of safety assurance (risk containment, risk assessment) criteria, definitions and schemes and corresponding legal frameworks applicable to laboratories in the various member states. The consortium EPIZONE (Network on epizootic disease diagnosis and control) improved research on preparedness, prevention, detection, and control of epizootics by improvement of excellence through collaboration. The DAISIE (Delivering Alien Invasive Species Inventories for Europe) project created an inventory of invasive species that threaten European terrestrial, freshwater and marine environments and provide the basis to prevent and control biological invasions through understanding of the biological, social, economic and other factors involved. The inventory is established using common definitions and criteria, and aims to cover all taxa known to be invasive, and all European countries, water bodies and seas. Where possible, the distribution of known invasions is presented graphically. The work also assesses the ecological, economic and health risks and impacts of biological invasions in Europe as well as indicators for early warning. In the frame of this project the European Alien Species Expertise Registry (http://daisie.ckff.si) was established. The Expertise Registry links and mobilises current expertise in biological invasions, to contribute knowledge and data to meet the requirements in dealing with invasive alien species. The European Expertise Registry contains details for individual experts with respect to taxonomic expertise, geographic units, and thematic areas.

PORT CHECK was a collaborative activity intended to deliver the tools and procedures to allow EU member state plant health competent laboratories and inspection services to perform molecular diagnostic assays "on-site" and at points of entry. The program was developed to evaluate real-time PCR assays for a number of key harmful organisms, and transfer these assays to field portable real-time PCR platforms which were originally developed for bio-warfare and bioterrorism applications. The project RIVIGENE (Genomic inventory, forensic markers, and assessment of potential therapeutic and vaccine targets for viruses relevant in biological crime and terrorism) aims to provide an integrated inventory of strains of security-relevant viruses, including their genome sequences, which are used in European and Russian BSL-4 high security laboratories. The project uses collated genome information to identify forensic genetic signatures of pathogen strains, and to distribute a simple and robust method of identifying these signatures to all laboratories of the consortium. The project permitted to build a network of laboratories of which each can identify all relevant virus strains in case of a release.

3. Future Research and Investments Needs

As envisaged at the EU Conference "Towards Future Challenges of Agricultural Research in Europe" (26–27 June 2007, Brussels) currently resources for agriculture–related research are in decline despite the emergence of new challenges. The FEG (Foresight Expert Group) recommended a new strategic framework for the planning and delivery of research, taking into consideration the complex dynamics operating between the domains of agriculture, food safety, food security, environmental preservation, water and land use preservation, societal impacts, rural sustainability, competitiveness, etc. However no specific actions were recommended with reference to biosecurity and, accordingly, the FP7 working programmes of 2007 included no calls on the issue.

The European Union, through its several Programmes, has the capability to increase preparedness within member countries through investments in infrastructures, to foster the development of national diagnostic laboratories and their networking, and to make investments in the basic and applied research needed to support the policies. The European Commission should recognize the need to confront the threats of crop and food biosecurity and financially support appropriate research. Further research and activities should include a strategic, long-range planning process to enhance the current surveillance infrastructure, to develop biosecurity capabilities in the agro-food sector, and to ensure that Europe has optimal capacity to anticipate, prevent, respond to, and recover from acts of agroterrorism.

All such programmes would be supported by reinforcing existing links between DG AGRI, DG RTD, DG ENTR, DG JLS and DG SANCO so as to deal with these issues with a multisectorial approach and including biosecurity as a priority of future working programmes.

Funding should deal with the following aspects:

- Molecular characterization, detection and identification of plant pathogens, plant disease epidemiology, disease modelling, risk assessment and prioritization, containment and eradication strategies, ground surveillance systems, and communication. The plant breeding community in Europe must be made aware of the bio-preparedness issue, to provide a basis for the appropriate use of pathogen resistance as a recovery mechanism from agroterrorism or biowarfare.

- Infrastructures. Upgrading existing diagnostic laboratories and building a network of such laboratories will be crucial. The fact that diagnostic laboratories currently differ significantly with respect to their capabilities and support poses a challenge to improve the diagnostic and communication infrastructure. Establishing a formalized network of laboratories will permit the European Union to a) establish an effective system linking diagnostic laboratories, universities and research centres, national and regional agencies, national expert laboratories; b) upgrade the diagnostic infrastructure present in each country; c) provide advanced training to diagnosticians; and d) develop data capture and analysis capabilities for the rapid identification of outbreaks. Laboratories belonging to the network will need special accreditation. The National Plant Diagnostic Network (NPDN), developed in the USA, provides a very good example (Stack et al., 2006).
- The ultimate goal should be the development of a European Centre for Biodefense to integrate regional resources for European security, drawing on resources from public, health, law enforcement and national security. Such a Centre should provide a unique, interdisciplinary capability to better defend against the full range of human, animal, and plant BSL-3 and BSL-4 biothreat agents. It should be developed by taking into account what already has been done in other countries, i.e. the United States.

Additional investments in training activities are needed. Training in crop and food biosecurity is much needed for those with responsibilities and interest in all sectors of crop agriculture, including extension specialists, students, crop consultants, regulators, and farm advisors at various levels. Crop and food biosecurity should become an important part of agricultural university *curricula* and researchers should be called to respect a code of professional conduct developed at the EU level, in accordance with other countries.

A robust programme (including infrastructures) for protecting EU crops and other plant resources will require significantly more personnel trained at all levels. Appropriate educational materials will be needed. Crop and food biosecurity should become important components of coursework for students graduating in plant health management as well as in food technology. Topics such as plant pathogen forensics should be tackled. This multidisciplinary field, which emerged as a new and important discipline for research, teaching and outreach, integrates traditional disciplines such as plant pathology and the specialized fields of forensic science and agricultural security. National agencies should establish opportunities for professional development and specialized training in agriculture and food protection, such as internships, fellowships, and other post-graduate opportunities to provide the needed European security professional workforce.

Moreover, long-term education and training programmes should be developed.

The next European Institute for Technology's Governing Board should be invited to explore the opportunity to design and select, among the other Knowledge and Innovation Communities (KICs), a partnership on plant science and biosecurity to undertake innovation activities, cutting-edge innovation driven research, education and training activities at the masters' and doctoral levels and dissemination of best practices. This should be done in coordination with EBN and existing network.

4. Networking

Groups of researchers having expertise and interest in crop and food biosecurity should be identified and the networks already existing should be integrated in order to build, at the European level, a solid critical mass capable to deal with the basic and applied aspects of research as well as with communication and training. The smaller networks already established in the field of crop and food biosecurity at the EU level should represent a core of interested parties from which a more inclusive one can be built, with special attention to the inclusion of researchers from the countries which recently joined the EU. The European Commission should play a major role in coordinating resources already available, promoting collaborations, and facilitating optimisation of processes at national, regional and local levels, in particular integrating efforts among regions. Unnecessary duplication of research should be avoided. On the contrary, a strong connection among labs should be sought. It will be extremely important to network researchers in this new field of expertise, at the European and international levels, to facilitate diagnosis, communication and training. In particular, it is important to develop and share new standards for diagnostic procedures, for the validation of such procedures, and for laboratory certification. Standards include microbial sequence analysis that differentiate at the subspecies or isolate level, reveal genetic modification of an otherwise known agent and facilitate decisions about the origin of the event (intentional, accidental or naturally occurring). In addition, state-of-the-art modelling and risk analysis developed in the different sectors will permit integration of epidemiology and economic risk assessment. It will be extremely important to cooperate with relevant authorities and agencies, such as EPPO, ICPP, EFSA.

Collaboration among government agencies, the commercial sector, international organizations and universities also is crucial: by combining a critical mass of knowledge and technology it will be possible to research and validate, assess and demonstrate innovative technological capabilities, operational concepts and procedures to assist in developing preventive and suppressive management of security challenges in the field of agriculture.

The public-private partnership is extremely important. Agricultural businesses, which have already very high standards of biosecurity, due to a very high degree of process control throughout their quarantine operations, can serve as examples to other industries. This process will also help companies in the new member-states to more quickly adopt the highest standards.

5. Information Sharing

Information sharing is a critical issue, as it requires a balance between confidentiality and public access, when or whether encryption is appropriate, and questions of database management and access. Apart from a need to respect information shared in confidence by growers, processors or other stakeholders, a policy of free and open exchange of information within EBN should be adopted. Information systems should be developed to bring all appropriate professionals and officials together in the development of an infrastructure to strengthen the EU agricultural industry through the enhancement of crop biosecurity.

Also requiring consideration is the issue of dual use information. A group of representatives from various EU countries might consider standards for such issues, focusing on the availability of agents, equipment and information that can be misused to produce bioweapons or used for terrorist purposes. Dual use problems require a set of norms and behaviours that will facilitate their management, without unnecessarily hampering scientific research.

A European Network would assist in the development of biosecurity and biosafety guidelines for publicly funded research and in the establishment of biostandards for publications and communication of sensitive issues. What constitutes sensitive information is still a topic of debate, necessitating better understanding of the term. Continued communication on the issue of publications among all stakeholders (publishers, security community, scientists, and lawmakers) is needed. The superior value of open communication between scientists compared to the value of social or economic security or *vice versa*, should be recognized. The dilemma consists in the choice between the free circulation of scientific information, that is the base of modern science, and the issue of maximum security, which would lead to suppression of public release of such information to avoid its misuse. The involvement of academic scientists and the novelty of the issue for the agricultural sector make the debate challenging. Decisions about dual use issues might best be addressed by a panel that includes scientists, security and law enforcement personnel, and policy-makers.

6. Need for EU Coordination

EU coordination, as well as coordination among agencies, is very much needed. Whereas in the United States numerous representatives of the Administration and Congress have publicly registered great concern regarding the threat of biological warfare, especially manifested in bioterrorism, in Europe crop and food biosecurity are not yet considered to be significant threats. The European Union should coordinate activities aimed at preparing European agricultural system at large to withstand a deliberate release of pathogen.

The European Union and all relevant European agencies should recognize the need to confront the threat and financially support appropriate research for fingerprinting high priority pathogens, detecting deliberate releases, developing rapid genetic-based diagnostic assays, epidemiology and risk prediction, and other scientific and technical approaches to reduce this risk. EU experience and ability to coordinate food defense work across borders, tested many times in the case of real-life events, is an important feature.

7. Outlook

It will be important to generate bio-preparedness through a comprehensive approach at the European and international level, also with the development of a European Network, which should include both public and private components.

Preparation for international cooperation in response to disease disasters at the regional or continental levels are poorly coordinated and cooperation is limited, although intergovernmental and international organisations have been advocating for years that emergency responses to infectious disease outbreaks should be planned for and prepared at the national level. National governments are responsible for contingency planning to protect the public; however, this responsibility needs to be broadened to encompass regional and international approaches.

Little public domain information is available on international coordinated responses to the deliberate introduction of biological pathogens. Terrorist events in the early 21st century have increased awareness of the risks and international cooperation in the field of agroterrorism and biopreparedness is at the beginning. The Center for Nonproliferation Studies, located at the Monterey Institute of International Studies, maintains a database of incidents involving sub-national actors and chemical, biological, radiological, or nuclear materials. The database lists 21 known incidents that could be classified as attacks against agricultural targets (http://cns.miis.edu/research/cbw/agprogs.htm).

Three international organisations are directly involved in human and animal health: the Food and Agriculture Organization of the United Nations (FAO), the

World Health Organization (WHO), and the World Organisation for Animal Health. None of the three organisations currently has a mandate to intervene, police, give opinions or assist in bioterrorism or agroterrorism events. However, during the May 2002 55th World Health Assembly held in Geneva, a resolution entitled 'Global public health response to natural occurrence, accidental release or deliberate use of biological and chemical agents or radionuclear material that affect health' was adopted. Under this resolution, the WHO established, under the leadership of the Communicable Disease, Surveillance and Response team, a series of initiatives addressing issues on the intentional release of pathogens; the aim was to improve international and national emergency preparedness to counter specific diseases and intoxications. In addition, the Global Outbreak Alert and Response Network (GOARN) was established to tap into the expertise of partner institutions and individuals and support the WHO's Alert and Response Operations group (ARO). At a WHO-organised meeting held at Lyon, France, in February 2005 on biological laboratory safety and biological laboratory security, Interpol, the International Criminal Police Organization, explained their role in investigating criminal acts of pathogen introduction, while highlighting their reliance on technical international inputs for understanding mechanisms of transmission risks and implications. A newly established unit funded by the Alfred P. Sloan Foundation is to focus on bioterrorism: that is, to raise awareness of the threat, develop police training programmes, strengthen efforts to enforce existing legislation, promote the development of new legislation, and encourage inter-agency cooperation in combating bioterrorism.

The Interim Commission on Phytosanitary Measures, which sets international standards for phytosanitary guidelines, oversees the International Plant Protection Convention (IPPC). The IPPC is a multilateral treaty for cooperation in plant protection that had its beginnings with the agreement by 12 countries to regulatory measures for grapevines under the Phylloxera Convention of Berne in 1881. This convention represented the first efforts at formalising international cooperation in plant protection and led to the recognition of the need to address other plant pests and enlist cooperation among all countries.

Global and trans-Atlantic partnerships to confront terrorism have existed for much of the 20th century, and events occurring early in the 21st century have increased society's awareness of potential threats to agricultural and food security. However, there is still a lack of solid, focused and well-resourced initiatives. The North Atlantic Treaty Organization (NATO), though originally created as a military pact in the aftermath of the Second World War, has established the Programme for Security Through Science, which offered support in 2005 for international collaboration on priority issues such as 'Defence Against Terrorism' and 'Countering Other Threats to Security'. This approach and other proposals from specific countries or institutions should be encouraged, when they apply to threats against the agricultural sector at large.

At this stage, it is extremely important to increase international cooperation on biodefense to protect agriculture and the food system worldwide. Strengthening and increasing convergence should be key elements of cooperation among EU and other countries such as USA, New Zealand, China and Israel, permitting the sharing of views, ideas, and experiences and enhancing preparedness all over the world. Additionally, international cooperation is critical to understand the effects of implementing new or enhanced food defense measures on various components of the agro-food industry.

Reference

Stack J, Cardewell K, Hammerschmidt R, Byrne J, Loria R, Snover-Clift K, Baldwin W, Wistler G, Beck H, Bostock R, Thomas C, Luke E. (2006) The national plant diagnostic network. Plant Dis. 90: 128–136.

EUROPEAN COMMISSION'S GREEN PAPER ON BIO-PREPAREDNESS

ISABELLE BÉNOLIEL
Directorate-General for Competition, European Commission
70, rue Joseph II, bureau 5/193, 1049 Brussels, Belgium

Abstract: The importance of developing a "biological all-hazards approach" which takes into consideration all risks (terrorist attacks, other intentional releases, accidents or naturally occurring disease outbreaks) is stressed relevant to European Union. As a consequence the creation of the European Centre for Disease Control (ECDC) sought to strengthen prevention and improve handling of outbreaks of existing and emerging diseases and biological health threats across Europe. Besides a Green Paper on Bio-Preparedness was adopted by the European Commission in July 2007 which is aimed at exploring ways to complement the framework already in place which seeks to ensure security and prevent deliberate criminal acts.

Keywords: European Commission, European Centre for Disease Control, pandemic, safety, cooperation

1. Introduction

Terrorists target our security, the values of our democratic societies and the basic rights and freedoms of our citizens – as the bomb attacks in Madrid and London in recent years as well as other unsuccessful attempts have clearly shown.

Although terrorists have mainly used explosives or improvised explosive devices, there is a real possibility that in future they could turn to non-conventional means, such as biological weapons or materials. Some of these materials have the capacity to infect thousands of people, contaminate soil, buildings and transport infrastructures, destroy agriculture and infect animal populations. These could affect food and animal feed at any stage in the food supply chain. Whilst the risk of a bioterrorist attack has been statistically low, the consequences of such an attack could be devastating.

M. L. Gullino et al. (eds.), *Crop Biosecurity*.

If a deliberate introduction of deadly pathogens or a naturally occurring disease outbreak were to occur in the European Union, or be imported from a third country, this could affect several Member States simultaneously, or spread across borders, having considerable economic and social impact.

2. All-hazards Approach

In our actions to prevent and deal effectively with potential attacks, we cannot focus solely on terrorism. The risks from dangerous biological materials and pathogens need to be reduced and preparedness enhanced through a "biological all-hazards approach" – generic preparedness within our overall crisis management capability. Such an approach takes into consideration all risks – terrorist attacks, other intentional releases, accidents or naturally occurring disease outbreaks.

3. Existing Instruments

A great deal of work has already been done to reduce the risks for EU citizens in the event of a bioterrorist attack or biological accident. Indeed, the European Union has developed structures which include instruments that have already proved effective in handling individual crises.

For instance, mechanisms are already in place for effective and rapid coordination in the event of human health crises, especially as regards agents carrying very serious threats, such as Anthrax, Botulism, Glanders, Haemorrhagic fever, Plague, Smallpox, Toxic syndromes or Tularemia.

Threats from the spread of communicable diseases have prompted the Commission to work to improve cooperation between the Member States. We have also developed systems and networks for disease prevention, early-warning and intervention.

The creation of the European Centre for Disease Control (ECDC), based in Stockholm, sought to strengthen prevention and improve handling of outbreaks of existing and emerging diseases and biological health threats across Europe. We have also developed legislation covering animal and plant health pathogens, as well as control and prevention mechanisms to ensure food safety, including strict hygiene rules, measures to reduce the incidence of salmonella and other zoonoses, the setting of microbiological criteria and animal by-product legislation.

Moreover, farm animals in the EU (cattle, sheep and goats) are subject to identification systems, which serve as a powerful tool to react to the discovery of threats, deliberate or otherwise. This is of great value in helping to isolate and eradicate diseases once discovered.

4. Crisis Management

The tools we have in place have consistently proved their effectiveness in the context of recent food and animal health alerts. For example, in 2006 and 2007 a number of meat scandals were detected with food expiry dates altered, and in some cases animal by-products were fraudulently entered into the food chain. In August last year, highly contaminated guar gum was found. Large scale traceability and recall operations in most Member States were immediately carried out as a result of these findings to isolate and remove the offending products from the market.

In relation to animal diseases, Avian Influenza (AI) has emerged as a new threat in recent years and Classical Swine Fever (CSF) and Foot and Mouth Disease (FMD) remain serious and recurring diseases. European instruments help to contain and eradicate these and other diseases whenever they arise.

5. Preparation for Biological Threats

The particular challenge posed by bioterrorism is the potential for multiple simultaneous outbreaks of highly contagious diseases.

Our defence measures were designed primarily with accidental or natural occurrences of threats to public health in mind. But given the increased threat of terrorism worldwide over recent years, the EU needs to consider the very real possibility that biological agents could be manipulated in order to carry out an attack. What pre-emptive measures might be taken to reduce the risks to human health in the event of an attack? How could the tools that have worked so far in response to other threats be used, tailored or improved?

The systems we have already, most notably a series of Rapid Alert Systems help the Commission and European Member States to issue timely alerts and exchange information on the situation with minimum delay. Moreover, the work done by the Commission and Member States to prepare coordinated responses to a possible pandemic, and the contingency plans that have been drawn up over the past few years, could also be put to use in the event of an attack which might threaten the health of a large proportion of the European population.

What is important now is to look at where improvements could be made in the systems in place, and how they could be adjusted to respond equally well in a bioterrorism attack as in natural or accidental crisis scenarios. This demands the effective coordination of all relevant actors – including national authorities responsible for risk prevention and response, human, animal and plant health; customs; civil protection and law enforcement authorities; the military; epidemiological and health communities, academic institutions; and bio-research institutes.

The Commission also plans to further develop its work in international fora and with key third countries. In this context, the sharing of best-practice and experience can be of invaluable mutual benefit.

6. Green Paper

A Green Paper on Bio-Preparedness was adopted by the European Commission in July 2007.

The aim of the paper is to explore ways to complement the framework already in place which seeks to ensure security and prevent deliberate criminal acts. It is another step in trying to improve our systems and to prepare for (or avoid if possible) any potential biological incident, accident or attack. The objective is to gain feedback from as broad a spectrum of society as possible, so that all angles are considered and taken into account in working to further improve our defences.

7. Objectives

The major issues of concern identified in the Green paper are:

- lack of awareness about the existing legislative framework;
- practical implementation of safety standards;
- existence and application of minimum security standards;
- potential to misuse research and researchers;
- deficits in European analytical capacity for reducing biological risks;
- lack of detection capabilities;
- need for improved multi-agency and multi-sectoral cooperation in both prevention and response.

These are the issues that directly affect the security of European citizens.

8. Consultation

We need to evaluate the mechanisms and frameworks that are already in place and how these are implemented, and then identify any shortcomings and propose specific actions where necessary. Stakeholders were invited to consider the options set out in the Green Paper; to indicate where they perceived current gaps and possible weaknesses; and to suggest what could and should be improved.

As the consultation ended relatively recently, we can only draw preliminary conclusions at this stage. We now have close to a hundred contributions.

This impressive response shows how much bio-preparedness has moved to the forefront of public concern. Stakeholders provided us with often extensive and high-quality comments, whether from academic institutions, industry associations, multinational companies, consultancies, and even third countries. Early indications show broad support for the “all-hazards approach”; our wish to avoid resorting to legislative means wherever possible; and our willingness to bring the public and private sectors together in pursuit of our common aims. Stakeholders rightfully reminded us that bio-terrorism ignores borders and invited us to enhance our cooperation at international level, both with third countries and international organisations.

They also pointed out that excessive actions, such as overly restrictive access to pathogens, could hinder much needed research on public health. They also expressed a wish to see greater coordination at EU level, such as with vaccine and drug banks and regarding disease surveillance.

The European Commission’s task in the coming months will be to analyse and summarise the broad spectrum of comments and suggestions that have been made.

9. Conclusion

A public synthesis of the outcome of this exercise will be adopted soon to identify the way forward. In parallel, the EU Member States adopted, on 6 December 2007, conclusions on bio-preparedness – which will also serve to give fresh impetus to this issue. This will help map out the agenda for further developments in the years to come.

APPENDIX

PROJECT WORKSHOPS AND SCIENTIST EXCHANGE

Workshops and scientist exchange represented a very important component of the project. It is well known that interactions among nations take place at different levels. Governmental officials generally communicate in formalized routes, and, because they represent their governments, they provide official viewpoints and approved positions. Scientists exchanges are typically less formal and take places through meetings, Workshops, Congresses. Another important way of communication and networking among laboratories takes place throughout exchange of scientists for short or long period. This permit scientists to interact on a day-to-day basis, exchanging views, ideas and sharing information.

Workshops were characterized by a small size, which permitted to build a strong connection among participants and to discuss in depth the many aspects and features of crop biosecurity. Workshops took place in Italy, Egypt and Israel, bringing together not only researchers but also policy makers, extension services and end users.

The exchange of scientists permitted to share information about diagnostic tools, web-assisted diagnosis, infrastructure organization, secure communication. Moreover, it helped the development of common research projects among the different Institutions.

Workshops and exchanges carried out are reported below.

Workshops

Workshop n. 1

Place and date: University of Torino, Torino (Italy), November 10th and 11th, 2005

Participants: James Stack (Kansas State University, US), Will Baldwin (Kansas State University, US), Jacqueline Fletcher (Oklahoma State University, US), Kellye Eversole (Oklahoma State University, US), Abraham Gamliel (Aro Volcani Center, Bet Dagan, Israel), Kamel Abd-Elsalam (Agricultural Research Centre, Giza, Egypt), Maria Lodovica Gullino (University of Torino, Italy), Angelo Garibaldi (University of Torino, Italy), Matias Pasquali (University of Torino, Italy), Daniela Minerdi (Agroinnova, University of Torino, Italy). Fausto Pedrazzini (NATO), Valeria Rizzo (Italian Ministry for the Environment, Land and Sea, Rome, Italy).

Topics covered: Overview of crop biosecurity in the different countries; plant forensics; molecular diagnostics.

Workshop n. 2
Place and date: Agricultural Research Centre, Giza (Egypt), March 18-20th, 2006
Participants: Kamel Abd-Elsalam (Agricultural Research Centre, Giza, Egypt), James Stack, (Kansas State University, US), Jacqueline Fletcher (Oklahoma State University, US).
Topics covered: Crop biosecurity; international cooperation; extension services in Egypt

Workshop n. 3
Place and date: ARO Volcani Center, Bet Dagan, Israel, March 23rd –27th, 2006.

Participants: Abraham Gamliel (ARO Volcani Institute, Bet Dagan, Israel), James Stack, (Kansas State University, US), Will G. Baldwin (Kansas State University, US), Jacqueline Fletcher (Oklahoma State University.US); Dan Levanon, (Ministry of Agriculture, Israel), Israel Ben Zeev and Edna Levi (Diagnostic Division, Plant Protection and Inspection Services, Israel); Yoram Kapulnik (ARO, Volcani Center, Israel); Dov Pruski (ARO, Volcani Center, Israel); Yephet Ben Yephet (ARO Volcani Center, Israel); Zeev Shmilovitz (ARO Volcani Center, Israel).
Topics covered: Israeli perspective for crop biosecurity; The need for biosecurity of natural and agricultural plant systems; Governmental, non-governmental and industry issues for international cooperation and collaboration ; General diagnostics and the Israeli Diagnostic Service; communications security and technology .

Exchange of Scientists

SURNAME AND NAME	ROLE	LABORATORY OF ORIGIN	HOSTED AT	PERIOD
Pasquali Matias	Post-doc	University of Torino	Kansas State University	April 18-28, 2005
Abd-Elsalam, Kamel	Researcher	Agricultural Research Institute, Giza, Egypt	University of Torino	November 10-18, 2005
Tinivella Federico	Ph.D. student	University of Torino	Kansas State University	April 3rd - June 30th , 2006

SUBJECT INDEX

H

I

K

L

M

N

R

S

T

U

Printed in the United States
112978LV00003BA/7-12/P

9 781402 084751